The Birth of Sensory Power

For Evelyn

The Birth of Sensory Power

Doing Things with Words, Numbers and Neurons

Engin Isin

EDINBURGH
University Press

Edinburgh University Press is one of the leading university presses in the UK. Publishing new research in the arts and humanities, EUP connects people and ideas to inspire creative thinking, open new perspectives and shape the world we live in. For more information, visit www.edinburghuniversitypress.com.

Edinburgh University Press Ltd
13 Infirmary Street
Edinburgh EH1 1LT

Typeset in 10.5/13 Bembo by
IDSUK (DataConnection) Ltd, and
Printed and bound in the UK using 100% renewable electricity by
CPI Group (UK) Ltd

A CIP record for this book is available from the British Library

ISBN 978-1-3995-3545-8 (hardback)
ISBN 978-1-3995-3547-2 (webready PDF)
ISBN 978-1-3995-3548-9 (epub)

EU Authorised Representative:
Easy Access System Europe
Mustamäe tee 50, 10621 Tallinn, Estonia
gpsr.requests@easproject.com

MIX
Paper | Supporting responsible forestry
FSC www.fsc.org FSC® C013604

This book is printed using paper from well-managed forests, recycling and other controlled sources

Contents

List of figures

These diagrams illustrate how nearly 8,000 summaries from approximately 200 academic books and articles were categorised and organised using MAXQDA, a qualitative data analysis application. These 200 items were derived from an initial reading of around 600 books and articles. The diagrams show high-level codes and their connections to the book's overall conceptual structure, which emerged simultaneously with the creation of these codes and their relationships. These diagrams serve to visualise the book's narrative progression.

Preface

The subject of this book is how twenty-first-century datasciences imagine the polity as a brain – as an intelligent, rational, self-organising, adaptive, instrumental, calculative and living organism – converging with neurosciences, which imagines the brain as a connected, affective, relational and plastic organism. The naming of datasciences and neurosciences as two forms of knowledge indicates that, although the disciplines such as data science and neuroscience are part of these forms, we are concerned with multiple concepts, practices and methods that assemble practical and theoretical fields to constitute forms of knowledge. These two forms of knowledge include diverse scientific and practical and discursive and non-discursive fields. The book is about bringing these two forms of knowledge into focus to give an account of their function in enabling a new form of power.

This new form of power now shapes our lives: the birth of sensory power. The ideologues, prophets and demagogues of sensory power would have us believe that sensory power is now the only power that matters in the twenty-first century and the future they claim to know as plenitude and singularity. The pervasive discourse on artificial intelligence, machine learning and the algorithms that create new autonomous technologies, from driverless cars to autonomous robots, self-generating cities and self-regulating polities, relentlessly remind us that a new age is upon us. The race to name this age as the surveillance, platform or algorithm age proceeds with many variations. We want to examine all this from a radically different but strangely familiar perspective on power. Yet our difference does not lie in viewing the same thing from a different perspective. We seek to bring an entirely different thing into view by naming it sensory power.

This is only possible writing in a style of thought on power from which we have become estranged. This style of thought arguably began with Baruch Spinoza (1632–77) in the 1660s and Friedrich Nietzsche (1844–1900) in the 1880s, continued by J. L. Austin (1911–60) in the 1950s, Gilles Deleuze (1925–95) in the 1960s and Michel Foucault (1926–84) in the 1970s and joined by Jacques Derrida (1930–2004) and Pierre Bourdieu (1930–2002) in the 1980s. This style of thought viewed power as a basic force of life that gives sense and value to it. As Ian Hacking (2002, 159–99) suggests, a style of thought is not a school, discourse or literature. A style of thought produces its own force, movements and qualities as heterogenous and irreducible elements that captivate, capture and activate its objects and subjects.

This book is as much about power as such and how power is exercised as it is about sensory power, which was arguably born in the 1930s, gathered pace in the 1980s and became sensible and intelligible in the twenty-first century. What does it mean to make this claim? Or, rather, what does this claim want to do? It was Nietzsche who invented a style of thought about the birth of things by creating several concepts to refine, develop and articulate it. *The Birth of Tragedy* (1872) had already signalled that there were not only different ways of seeing things differently but there were different things that thought can bring into being.

Nietzsche seeks the origins of Greek tragedy because he believes it contains two styles of thought as drives that combine to create tragedy: Apolline and Dionysiac. If Nietzsche thinks that these are figures, let alone opposites, he will awaken himself to see that they depend on each other as drives. Tragedy is a pairing of the two, despite their conflicts. Apolline drives dreaminess; Dionysiac drives drunkenness. To understand them, we need to separate them and then see how they are bound together in their function as gods, Apollo and Dionysus. Apollo's serenity is opposed to the restlessness of Dionysius. Nietzsche believes that these gods are artistic forces that erupt from the earth itself. But to present these two as the drives of a style of thought is only the beginning: Nietzsche must understand the degree and level to which these drives have developed in a people. Nietzsche thinks that the Apollonian Greeks must have been astonished when the Dionysian Greeks were born within them. 'With an astonishment enlarged by the added horror of realising that all this was not so foreign to them after all, indeed that their Apolline consciousness only hid this Dionysiac world from them like a veil' (Nietzsche [1872] 1999, 21). To understand this moment of birth, Nietzsche would 'need to dismantle the artful edifice of Apolline culture stone by stone, as it were, until we catch sight of the foundations on which it rests' (Nietzsche [1872] 1999, 22). We must reveal the deeds that built this edifice and the needs that drove these deeds. If we were a spectator rather than a genealogist, as Nietzsche ([1887] 2007) would eventually name this style of thought, we would seek what we already know. But genealogy is not a methodology; it is a strategy. It involves listening to what the needs and deeds say rather than turning away. We may then understand that, 'in order to be able to live, the Greeks were obliged, by the most profound compulsion, to create these gods' (Nietzsche [1872] 1999, 23). Nietzsche asks: '[H]ow else could that people have borne existence given their extreme sensitivity, their stormy desires, their unique gift for suffering, if that same existence had not been shown to them in their gods, suffused with a higher glory?' (Nietzsche [1872] 1999, 23–4).

In just a few paragraphs Nietzsche outlines the principles of genealogy as a strategy for revealing the ways in which edifices are built with words and deeds driven by affects. That the birth *of* tragedy involves two drives implicated in each other to the astonishment of a people who experience their birth *as* tragedy is the principle of genealogy that inspires *The Birth of Sensory Power*.

Michel Foucault explores several uses of genealogy, as Nietzsche ([1887] 2007) came to describe his way of seeing things differently. Foucault's work on parrhesia in ancient Greek and early Roman speech acts reveals his understanding of genealogy as strategy of speech acts. Foucault first identifies how parrhesia functions as openness and sincerity in speech. The parrhesiast, or the one who performs parrhesia, is expected to speak without hiding anything and to use direct and clear language. Foucault then distinguishes between the active and reactive uses of parrhesia. While the active use of parrhesia involves speaking the truth with sincerity and conviction, the reactive use is characterised by speaking without discretion or care. The concept of parrhesia as a speech act establishes a direct relationship between the speaker and the truth. There is danger in parrhesia, requiring the courage to speak the truth. For Foucault, then, parrhesia is not simply about speaking the truth but a confrontation with oneself and others that is both solidaristic and agonistic but always in a situation where the speaker is seeking justice. This makes parrhesia not only dangerous but also obligatory. The parrhesiast is seen as exercising their freedom to choose truth over lies, and to fulfil their obligation to confront others. Foucault's parrhesia as speech acts concerns making things happen or appear in words and deeds; it retains its resonance with the will to power as Nietzsche would understand it.

Hannah Arendt ([1958] 2018) comes to genealogy as strategy with the concept of 'acts' to signify beginning something new into the world whose outcomes are unpredictable by exploring danger as obligation. If humans were repetitions of the same whose essence was predictable we would not need the concept of acts (Arendt [1958] 2018, 8). Arendt insists that each human shares the condition of natality, which cultivates the drive to begin something new because we are born with the capacity to come into the world as new (Arendt [1958] 2018, 9). Our acts reveal this capacity and disclose each human in the presence of and in relation to others (Arendt [1958] 2018, 176). This disclosure of ourselves is something we can neither anticipate nor determine. To act then means to take initiative, to begin and to set something new into motion. Because humans are newcomers and beginners by virtue of birth, they are always prompted into action, into disclosing themselves to others. Arendt calls this capacity, or what enables human beings to make things happen, a 'miracle' – not because it is mystical but because it is unexpected and unpredicted. When humans perform acts, they are capable of enacting the improbable and unpredictable (Arendt [1955] 2005, 114). Each human is capable of performing the unexpected and it is only through this performance that each human being is able to disclose themselves as human. We can perform what is currently unimaginable. With each birth something uniquely new comes into the world (Arendt [1958] 2018, 177–8). What an act reveals or discloses is not a being that already exists but a being that emerges through this disclosure or revelation. That is why our rights are obligations of the other to us and the rights of others are our obligations to them; our acts are dangerous but necessary.

The resonance of Arendt and Foucault with J. L. Austin is remarkable but recognisable. If for Austin speech acts mean that 'in' and 'by' saying something we are doing something, the idea that speech is not only a description (constative) but also an act (performative) is a radically different way of thinking about not only speaking but also doing things in or by speaking. For Austin, even constative speech acts as statements describing a situation cannot remain without effects. It is in this sense that Austin makes us think about ourselves as those beings who always know how to do things with words, even though our chances of accomplishing things we set out to do remain precarious if dangerous, since our words will occasionally (perhaps more so than we would like) misfire and make things happen that we did not intend.

The style of thought that emerges from Spinoza, Nietzsche, Foucault, Deleuze, Derrida and Austin, among others, is not only about speech acts as an obligation to speak truth of a thing but also about bringing things into being as a new relation between words and things.

We have been living through the invention of sensory power from the 1980s to the 2020s, which has been perhaps its most intense period since the 1930s. The phrase 'living through' means that, as subjects of sensory power, we have been affected and effected by its strategies and technologies. We are realising that we are interpellated or called upon to perform ourselves in radically different modes through platforms, algorithms, devices and sensors; that we are becoming integrated into learning machines whose operative principles are concealed; and that a transformed game – the imitation game – is upon us. With astonishment heightened by the realisation that this game is not so strange after all, it appears crucial to rework a style of thought on power to deconstruct the game, as if by taking it apart, until we catch a glimpse of the code upon which it rests, and perhaps even to put it together differently.

Introduction: the games we play

The book explores the convergence between neurosciences and datasciences in the production of an imaginary of polities as learning machines: an imaginary that institutes various concepts, practices and methods, deterritorialises and reterritorialises cities, states and empires, produces autopoietic subjects and resignifies the citizen. To explore this imaginary is urgent: capturing, storing, analysing and visualising sensory data about emotions, affects, moods, activities and movements of people by datasciences increasingly transform polities to function in the way in which modern neurosciences understand the development of the brain. There are questions about how twenty-first-century polities are being modelled after a certain understanding of the brain as revealed by new forms of knowledge (neurosciences and datasciences) and forms of power (sovereign, disciplinary, regulative and sensory).

While a history of knowledge and power regimes for producing new type of polities and a citizen-subject is the background, we are primarily interested in naming a new form of power – sensory power – as revealed by twenty-first-century developments. The imaginary of the twenty-first-century polity is both an effect and affect of this new form of power that combines elements from neurosciences and datasciences to create a new citizen-subject and a new polity. We want to know how this imaginary has been instituted, by which agents and actors, and for which interests. From neurosciences this imaginary draws an interpretation of the brain as a malleable and changeable organ, its plasticity. Sensory power attempts to govern not only through the body but also through the brain, or, rather, the object that embodies the body and the brain. From datasciences it draws a sensory infrastructure built from devices and networks in which the body-brain is deeply connected. Sensory power mobilises citizen-subjects for governing themselves as autopoietic subjects and the polity functioning as a brain-body. The citizen-subject and polity combine both human and non-human assemblages to accomplish certain goals or objectives. The polities such as cities, empires, colonies, states, nations and federations have affinities and overlaps with these polities but polities as learning machines now traverse existing populations, territories and peoples.

The long history of power

The book explores the convergence of these pivotal developments in the formation of contemporary polities. It explores the emergence of an imaginary of the citizen against the background of new forms of power, dating back to the origins of polities approximately 6,000 years ago. Instead of presenting this background as a linear historical progression, we identify ruptures and reversals where polities were organised around specific games of power, namely kingship, citizenship, population and imitation.[1] While we can identify historical moments when each game became operational – 6,000, 2,500, 250 and 100 years ago, respectively – each moment has also meant the resignification and reorganisation of the others.

The emergence of datasciences with distinct concepts, practices and methods in the twenty-first century, especially precipitating massive amounts of data collected through ubiquitous devices, constitutes a transformative moment in how power is exercised and it is our beginning point. The combination of geolocation, geosensing and geofencing have increasingly involved sensors in mobile devices to forecast, predict, anticipate and modulate behaviour of both people and things (Desrosières 2016). With sensory data captured from devices, datasciences are increasingly making claims to govern people's affects, moods, feelings and emotions (Gabrys 2019). We have already witnessed their effects in twenty-first-century polities (influencing voter behaviour), commerce (influencing consumer behaviour) and culture (influencing social behaviour). Yet their deeper affects and effects may well be concealed in experiments in new polities as learning machines. The history of these experiments and their articulation require a long view on power.

The emergence of neurosciences since the middle of the twentieth century, constituting the brain as an object with concepts, practices and methods, was a transformative moment. Understanding how neurosciences converged with datasciences generated a decisive analogy between the machine and the brain also requires a long view on power. To put it emphatically, if the eighteenth- to twentieth-century imaginary of the brain as an organ ensconced in the skull and developed early in isolation, it has been displaced since the mid-twentieth century by the imaginary of the plasticity and connected brain.[2] This imaginary was only possible with the convergence between datasciences and neurosciences and the emergence of the brain–machine analogy – the imitation game. This has not been without opposition as various criticisms of biological determinism and neurological reductionism also soon followed (Damasio 2006). There have been different emphases on exactly what it means to understand the brain, ranging from assertions of 'who we are' to 'you are your brain' and to 'how our brains become who we are' (LeDoux 2002).

Nonetheless, in the twenty-first century, the imaginary of the brain has gradually changed its focus from a static organ to a complex biological, social and cultural body-organ (Pitts-Taylor 2016). While neurosciences first intensified the nature versus culture debate and the struggle between humanities and sciences, it seems more recently to have established a dominant position where the brain is seen as a continuously developing connected organ whose study comprises both humanities and sciences. For some this dialogue led to 'neuromania', where, by simply adding 'neuro' to their names, many fields and disciplines began making claims under descriptions such as neurolaw, neuroeconomics, neuropolitics and neurourbanism (Rose and Abi-Rached 2013). For others, the development of neurosciences also meant an exuberance and vitality that is encouraging for establishing a dialogue between humanities and sciences (Tougaw 2018). Yet the convergence of neurosciences with datasciences also precipitated various experiments in creating or recreating the polity and the citizen, which is the focus of this book.

If indeed these two forms of knowledge – datasciences and neurosciences – converge in consequential yet less spectacular ways for the reorganisation of polities, we want to develop analytics of power appropriate to understanding these forms of knowledge (Mattern 2021). The sensory projects driven by datasciences and neurosciences throughout the planet earth and its orbit have increasingly imagined the polity as a system governing itself through continuous responsiveness from citizens with embodied sensors in their devices (Kitchin, Lauriault, and McArdle 2018). We want to understand the history of how this happened.

As an imaginary of the brain as a connected entity emerged, long histories of both datasciences and neurosciences have shattered the dominant narratives of the development of nervous systems in their social and cultural conditions (Smail 2008). There is now an imaginary where we see the domestication of plants and animals as associated with the domestication of humans – our brains and our bodies. Often stated as how culture might have influenced human physiology and how neurochemicals associated with affects (emotions, feelings, moods) have histories embodied in societies and polities, we are increasingly seeing the development of the brain as simultaneously biological and social, and we are deploying these insights in governing ourselves through creating polities that imitate the new imaginary of the brain.

How to do things with power: words, numbers, neurons

A long history of power can be called 'historical ontology'.[3] If we want to insist that beings such as brains and machines have a history, then historical ontology is an apt name for exploring how these beings come to be. Following Hacking, as much as we are interested in neurosciences and datasciences, we are keenly focused on the deployment of strategies and technologies for governing the movement and conduct of peoples and polities through concepts, practices and methods. We shall explore in detail the interconnected developments of concepts, practices and methods in both datasciences and neurosciences, but for now it is important to highlight two forces where these strategies and technologies become sensible and intelligible in/on polities: projects and professions. The former involves initiatives to deliver 'solutions' to the problems of making polities efficient, healthy, functional and beautiful. The latter involves people organising expertise to articulate the problems to which projects deliver solutions. As Pierre Bourdieu would say, the competitive struggles among the professions organise practices where knowledges such as neurosciences and datasciences deliver projects as solutions (Bourdieu [1984] 1988). The process involves the accumulation of forms of capital (symbolic, social, economic, cultural) and the professions drive their conversion through projects. These practices produce the imaginary of the polity and of citizen through symbolic and material objects scattered through organisations (institutions, corporations, governments). It is this assemblage of organisations, institutions, projects, professions and various forms of capital, forms of knowledge and forms of power that constitutes the objects such as machines and brains that historical ontology investigates.

The professions and projects operate with specialised languages. To understand language as social activity rather than only means of communication means investigating how people do things with words in ordinary language. J. L. Austin (1962; [1961] 1970) made an immense contribution to the relationship between words and things.[4]

Although for Austin a speech act often involves doing things with words, he is critical of the dominant view that a speech act is merely a description. Austin calls this descriptive element of speech acts constatives. When we use statements to describe a situation we are performing a constative speech act. As truth claims, such statements can be verified or falsified. As a constative speech act, it is a statement that signifies a meaning. By contrast, when a statement either warns about something or urges someone to do something, it moves from being a statement to being an utterance. The latter accomplishes an act by its *force*. Austin says that there are many verbs in the English language that can be classified according to these effects of meaning and force. Austin uses three connectives to classify speech acts: 'of', 'in' and 'by'. For 'of', Austin says that the act *of* saying something is a locutionary act. This is a speech act whose *meaning* calls forth a truth versus false distinction and requires verification or falsification. The effect or consequence of a locutionary act – a constative – is to produce or fail to produce a meaningful description of a situation. For 'in', as *in* saying something we may be doing something, Austin says it produces an illocutionary act. This is a speech act whose *force* creates a *potential* effect in a situation that it seeks to describe. What it invokes is not verification or falsification but whether there is an uptake or whether it works. To put it another way, whether an utterance is successful (felicitous) or unsuccessful (infelicitous) is determined by the force it generates to make things happen. Finally, doing something *by* saying something is a perlocutionary act. This is a speech act that must have an affect or effect to be actualised. Like an illocutionary act, a perlocutionary act invokes an evaluation along felicitous or infelicitous lines rather than true or false. By saying something, people accomplish something. Thus, 'of' saying something has *meaning* (locutionary acts), whereas 'in' or 'by' saying something has *force* (illocutionary and perlocutionary acts). Locutionary speech acts are propositional. Illocutionary and perlocutionary acts are performative.

The distinctions between meaning and force, between statement and utterance, and between constative and performative are key to understanding how projects and professions makes things happen with specialised languages. By advancing the idea that speech is not only a description (constative) but also an act (performative), Austin ushers in a radically different way of thinking about not only speaking and writing but also doing things in or by speaking and writing (Cavell 1994, 81).

Austin knows that words will never be enough to accomplish an act, despite recognising that without words at some stage it is difficult to see how any act can be accomplished at all (Austin 1962, 12). So, although his examples are from speech, his interest is how words perform acts: 'to say something is to do something; or in which by saying or in saying something we are doing something' (1962, 87). Austin also recognises that nonverbal forms of speech such as bodily gestures and movements as well as visual and aural forms are always involved in the accomplishment of an act, especially in a perlocutionary act. So Austin may not mention bodies much, if at all, but bodies and their movements are implicit in speech that acts. To put it differently, speech or writing cannot act without bodies. Although all of Austin's examples are in the first person, he is not fooled by the speaking sovereign subject who is the master of their speech situation. On the contrary, as the illocutionary and perlocutionary acts indicate, in acting there are always infelicitous situations. Cavell notes that Austin recognises that acts will *also* occur 'unintentionally, unwillingly, involuntarily, insincerely, unthinkingly, inadvertently, heedlessly, carelessly, under duress, under the influence, out of contempt, out of pity, by mistake, by accident [and so on]' (Cavell 1994, 148). To put it differently, Austin's concern with infelicitous effects is not a regret on his part but a recognition that speech does not only act; it also can fail to act or fail to act in ways anticipated.

All this would lead Austin to say investigating an act would require fieldwork – we could say historical ontology or genealogical investigations – though he admits that is not what he is doing (Austin 1962, 148). Those who wish to explore how conventions in, of and by which people act must undertake genealogies on the meaning and force of statements that perform acts and make things happen. Both the meaning of statements and force of utterances become possible by their appropriateness to the situation, which means an understanding of conventions governing a speech act situation. To judge the situation, people must understand and perform in relation to the conventions governing what is sayable and doable in that place and time. The people in action will have an understanding of not only the situation but also the appropriateness of what can and must be said and not said in that situation (Austin 1962, 8–9, 13, 26, 29, 34, 81–2).

This is why Pierre Bourdieu (1991, 125) will insist that words themselves do not have inherent *meaning* or *force* but acquire them in social situations. For Bourdieu, speech acts are social acts and draw their force – illocutionary force, if you like – not from linguistic apparatuses that govern them but from social apparatuses that make them possible. The things that render a subject capable of accomplishing an act are precisely the social apparatuses that guarantee its institution (Bourdieu 1991, 125–6). Bourdieu clearly does not deny the power of words. But such power is nothing other than the power of the speaker as a social subject (107). However, do speaking subjects *merely* follow conventions (as Bourdieu seems to think)? The key issue in speech acts becomes whether, and if so to what extent, what is sayable and articulable follows or exceeds social conventions that govern a situation.

For Austin, a general consideration is that, although a people will rely on conventions to ascertain the meaning and force of their speech acts, neither the meaning nor its force will ever be controllable by them. So, while a convention is a necessary condition of an act, almost equally it is also a necessary condition of its misfire – one of the 'ills that all action is heir to' (Austin 1962, 16, 25, 27, 105). That it requires conventions to accomplish an act and that it also provokes their transgression turns out to be a significant discovery by Austin. Judith Butler makes use of this in the theorising of gender by drawing our attention to the citation, repetition and resignification of a convention and how these produce subjects of both submission *and* subversion. If a convention is to be cited to accomplish an act, a repetition of certain norms will be necessary. Yet each repetition will bring new circumstances to bear on the act, so much so that it is a resignification – a new deployment of a convention (Butler 1990, 173, 177–9; 2004, 218–31). But Austin reserves the transgression, or subversion, if you like, of a convention only for perlocutionary acts (Austin 1962, 120–1). While illocutionary acts will cite and iterate conventions to enact their performative force, perlocutionary acts will derive their force from the unconventional, undecidable and unpredictable effects.

If for Austin speech acts mean that 'in' and 'by' saying something we are doing something, the idea that a speech act is not only a description (constative) but also an act (performative) ushers in a radically different way of thinking about not only speaking but also doing things in or by speaking. For Austin, even constative speech acts that are statements describing a situation cannot remain without affects and effects. It is in this sense that Austin makes us think about ourselves as those beings who always know how to do things with words, even though our chances of accomplishing things we set out to do remain precarious since our words will occasionally (perhaps more so than we would like) misfire and accomplish things we did not intend.

What about saying something in or by doing something? Or, to put it differently, as Bruno Latour (2000) puts it, how things may perform words? This question may appear redundant since Austin is said to have questioned the difference between saying and doing. But Austin never considers speech situations where things perform words. It was Latour who raised the question of how we may accomplish words in or by doing things – hence reversing Austin's phrasing. Latour's reminder that actions can accomplish words is crucial, as we will see in exploring how to do things with numbers and neurons.

Now Austin does not explain what is an action that accomplishes an act. He implicitly recognises that action is a bodily movement that accomplishes something but he does not explicitly recognise that to perform an act involves bodily movements. We may raise our hand, for example, in a particular way to indicate we are taking an oath – an act of promising. Here, raising a hand is an action, but the conventions governing the situation indicate that this action is performing an act of promise. An action is a bodily movement that accomplishes something. Strangely, Austin says that performing illocutionary acts does not require action. This is because the illocutionary force of an act consists in its potential. This is the principal reason we need to investigate not only things done in or by speaking but also things said in or by doing things. If we are interested in doing things not only with words but also with numbers and neurons, we have to think about what it is that people or things are doing with numbers and neurons. They are doing things with power. We use the phrase 'how to do things with power' or 'doing things with power' as a means to think about power as such.[5]

We will see later that power as such does not exist; or, rather, its effects and affects can only be shown through the forms through which 'it' functions. We will discuss sovereign, disciplinary and regulative forms of power and their historical specificities. For now, let us highlight how a new form of power – sensory power – was made sensible and intelligible during the coronavirus pandemic and illustrate why taking a long view on power is necessary.

The coronavirus pandemic as an event

The coronavirus pandemic as an event may have made not only sensory power sensible and intelligible but also sharpened our sense of sovereign, disciplinary and regulative forms of power.[6] The responses to world events such as pandemics, epidemics, wars and disasters not only accelerate the latent developments in social, economic, cultural and political practices, technologies and arrangements but also reassemble and reorganise them. Such events may not be understood well during their happening as their affects and effects take long periods to crystallise. Yet such events also destabilise our habits and styles of thinking that provide openings to reevaluate those habits and styles. Throughout the coronavirus pandemic, national and international governing authorities across the world made life-and-death decisions for both citizens and non-citizens. It will be debated for some time how certain deaths could have been avoided, whether people could have been better protected, and if a sensible balance between health and economy could have been maintained. It will also be debated how, by protecting 'their' citizens, powerful authorities may not only have exacerbated or even entrenched global inequality but also prolonged the end of the pandemic – a tragedy to be named in the future.

Throughout the pandemic national and international governing authorities awarded, cajoled, cared, charged, confined, counted, cured, detained, encouraged, entertained, deceived, dissuaded, fined, furloughed, imprisoned, informed, killed, misinformed, notified, ordered, persuaded, pleaded, punished, quarantined, searched, stopped, traced, tracked and vaccinated people. This is quite a range of government performatives, and their intensity may have been unprecedented in scope and scale. The list includes only a fraction of ways of exercising sovereign, disciplinary and regulative power. Yet people are not always obedient or submissive, and they resisted differently in different times and places and exercised power in collective ways. The initial obedience about confinement, for example, gave way to some dissent. There was always some dissent about restrictions on movement, about coverings and about vaccination. Conversely, the initial dissent about vaccination gave way to submission. It would be an exhausting list if we were to add all the ways in which power was exercised by national and international governing authorities and how people both embodied and confronted them.

Is there a way to make sense of power relations between polities and peoples that the pandemic made sensible and intelligible without reproducing a hierarchical or flat idea of power? Can we group all these ways of exercising power into various and diffuse forms, reflect on their historical development, and then ask whether this exercise can provide insights about power in the present? Would such an exercise help us better understand the birth of a new form of power in the present? These are some of the questions that the coronavirus pandemic has provoked.

We have mentioned just above that there are various forms of power through which we can identify as such power. It was Michel Foucault who named three different forms of power by virtue of their distinct logics and historical development. These are sovereign, disciplinary and regulative forms of power. These forms have developed from the seventeenth to twentieth centuries in Europe and outside Europe. The logic of sovereign power is that it demands obedience. It is an asymmetrical form of power with often deadly consequences. Although we witness the birth of sovereign power in the seventeenth century it has a deeper history. Moreover, its exercise can still be found in modern institutions where sovereign power performs its interdictions. In other words, sovereign power still exists but the ways in which it is exercised have gone through numerous transformations, especially with the birth of subsequent forms of power. The logic of disciplinary power, whose origins can be witnessed in the eighteenth century, involves governing people first by subjection and then by submission. If it is imperative to the state that its people be healthy, then its exercise will involve pressure, persuasion and, if necessary, punishment. The logic of regulative power made things more complicated since its birth in the nineteenth century. Its logic functions by developing indicators and metrics to measure the health of a people as a whole and invents technologies for governing each person to enhance these indicators, thereby making people healthy.

There are two important things to bear in mind about these three forms of power. First, we speak of the birth of a form of power in the sense of its strategies and technologies coming together in an effective and affective manner at a certain time and place. The technologies that constitute a form of power may take a long time (and many places) to develop. The birth refers to a time and place of their effective and affective assemblage so that we can see, observe, name, feel and speak about it. Second, the birth of a form of power does not mean the death of another. What increasingly complicates modern politics is that all three forms – and, as we shall see below, the fourth that the pandemic made sensible and intelligible – became forms for governing people. These two points are important for the argument that the pandemic made a form of power sensible and intelligible. As we shall see, however, its strategies and technologies have been developing since the 1980s and especially since the 1930s.

If all forms of power govern people through technologies, some historical examples may help: sovereign power governs through colonies, dominions, empires and states and invents armies, borders, cadastres, cartography and partitions; disciplinary power governs through, camps, cities, hospitals, factories, prisons, schools and workhouses and designs buildings, policing, walls, fences, guards and gates; and regulative power governs through counting, recording, enumeration, metrics, tabulation and attributes such as class, gender and race.

The long list of government performatives above can only be accomplished by diverse strategies and technologies of power. Consider how three forms of power have required specific technologies during the pandemic: sovereign power requires obedience (lockdown, curfew, confinement); disciplinary power extracts submission (sacrifice, distancing, isolation, quarantine, hygiene); and regulative power tracks the relation between individual conduct and collective health (infection, transmission, mortality, recovery and immunity, all expressed in rates).

It is necessary to recognise that sovereign power has a long history that goes back to the development of the first cities, empires and states about 6,000 years ago. But the development of early modern European empires and states in the seventeenth century by consolidating wealth, territory and people in novel ways marks its birth in the way in which we recognise it. Straddling the seventeenth and eighteenth centuries, the birth of sovereign power is associated with, on the one hand, the birth of modern empires and, on the other, state apparatuses with which they were governed. In this period, the concern of sovereign power increasingly became the health and wealth of subject peoples. European empires were built by accumulating peoples through slavery, colonising indigenous peoples and settling colonies. If each of these apparatuses required extracting obedience from subject peoples, they also precipitated searching for more effective and efficient ways of governing both colonial and metropole peoples. If the key objects of government were territories governed through technologies of settlement, deportation and dispossession, key forms of knowledge also developed on the wealth and health of subject peoples known as political arithmetic. The accumulation of mercantile capital would have been inconceivable without the transatlantic slave trade of African peoples into colonial settlements, the subjugation of indigenous peoples, and deportations of dangerous peoples by forcing them into becoming settler colonists. If governing metropoles meant subjugating 'dangerous' populations by cruelty, it meant governing colonies by mass occupation, displacement and dispossession.

With the effective closing of state borders and mobility restrictions during the pandemic, some imagined that this was a return of sovereign power. It has, however, always been present, but its exercise involves a different assemblage of technologies of power. Lockdown, curfew, confinement, regulation of movement, border controls and overall restrictions on the mobility of subject peoples are the most common technologies that sovereign power developed over a long period. What we have experienced is new ways of assembling technologies of sovereign power during the coronavirus pandemic. Some states were praised for acting most swiftly in shutting borders and some for introducing internal borders by enclosing entire neighbourhoods, cities and regions. Both external and internal borders were subjected to immediate controls on all continents, though with varying intensities across different states. That so many people reacted with surprise illustrates how technologies of sovereign power become so entrenched in our lives over time to have become nonsensible and unintelligible by inhabiting our bodies. It also illustrates how the privileged citizens of powerful states do not normally experience the brutal and cruel effects of sovereign power, unlike non-citizens and non-state peoples such as refugees or indigenous peoples.

We thus ought not to conflate invisibility with inexistence. Nor should we be surprised by the widespread obedience that sovereign power has extracted despite occasional and scattered protests primarily in the United States but also in Europe questioning restrictions on movement. A most powerful lesson from the pandemic is probably that sovereign power is entangled with other forms of power, from which it draws strength but also needs to be analytically separated. Unlike the seventeenth-, eighteenth- or nineteenth-century variants, sovereign power in the twenty-first century could not have functioned without relying on disciplinary and regulative forms of power, which undoubtedly contribute to its persistence.

From the eighteenth to nineteenth centuries, we witness the emergence of a new form of power: the discipline of capacities and desires. The key object is the body but in a different way. It is true that sovereign power is also exercised on the human body: cruelties from branding black bodies to spectacles of torture and deportations with fatal consequences are widespread technologies. Governing people required exercising sovereign power as the right to decide over the life and death of bodies.

Yet, the logic of disciplinary power is submission that makes the body useful and healthy. It may have started in barracks (soldiers) and on ships (slaves) but it invented technologies of power and forms of knowledge that combined to create optimised bodies. Over the next three centuries disciplinary power produced clinics, prisons, hospitals, schools, workhouses, camps and eventually gyms, malls, studios and other assemblages where forms of knowledge were brought to bear on people governing their bodies themselves.

How have we experienced disciplinary power during the pandemic? Just consider how we have collectively become experts in protecting our bodies and the bodies of others. We have not only followed daily what sciences have discovered about the virus and its modes of infection but also internalised injunctions and admonishments on how to conduct ourselves safely for others. We have been advised to sacrifice everyday activities by isolating to save ourselves, others and public health and care systems. We have developed, in an astonishingly short period of time, new forms of conduct by protecting ourselves and others in physical distancing, covering our faces and regulating our contacts. We have developed astoundingly ritualised hygiene practices of disinfecting ourselves. We have exercised all these forms of submission that disciplinary power calls for our own and each other's health and safety. If we followed the rules of confinement imposed by sovereign power obediently, we followed the rules of safety called for by disciplinary power submissively. What the pandemic has rendered sensible and intelligible is that we experience these two forms of power simultaneously. We, our bodies, recognised how these two forms of power – sovereign and disciplinary – depend on each other and work together. Under the pandemic circumstances they become revealed. Without a hint of irony, for those who needed help, practical guidance was offered on how to relearn socialising after the confinement.

The confinement of people by partial or total lockdown was certainly governed by punishment: fines, charges, attestations, permissions and identity cards were mobilised to separate those who were successfully responding to sovereign power by exercising discipline from those who were not. The logic behind this obedience is that submission to discipline was good not only for us and others but also for yet another: a collective yet invisible body whose health depended on us and others. What is this collective body that we have been protecting and that has been presented to us by strange metrics we had heard so little about before such as herd immunity and infection rates?

From the nineteenth to the twentieth centuries, we witness the emergence of a power that is at once collectivising and individualising. It is regulative power exercised on a people, a collective body whose health is measured by fertility, mortality, longevity and reproductivity. By simultaneously individualising and specifying, it is directed towards the performances of individual bodies, with attention to the processes of life. Regulative power bridges sovereign power and disciplinary power by effectively linking individual behaviour to the behaviour of a people focusing on calibrating or adjusting them to each other. Each form of power depends on the other but now regulative power functions most effectively as a positive rather than negative force. What regulative power mobilises is the prescription of appropriate forms of conduct for bodies that are necessary for, or conducive to, the functioning of a people's health and wealth. More importantly, regulative power regulates a people not by admonishing or punishing for non-compliance (though that relationship between sovereign power and disciplinary power continues to function) but by persuading, guiding, nudging and cajoling bodies, asserting that their health and wealth derive from it. Bodies conduct themselves as responsible subjects for their own and for the common good for the health of a people to which they belong.

Above all, the coronavirus pandemic made us recognise that playing with power requires performing all its three forms in their entanglements. Yet, and so we think, the pandemic may have made a fourth form of power sensible and intelligible. If so, this makes theorising power more complicated.

How can we claim that we are experiencing the birth of a new form of power? We are now aware that we are living in a planet in which surveillance (of children, citizens, clients, consumers, criminals, customers, friends, inmates, lovers, non-citizens, patients, spouses, students, terrorists, workers) has become ubiquitous and pervasive. The devices from wearables to appliances and from computers to cars are being traced along with everybody's activities tracked with them. The insatiable appetite of authorities – states, corporations, militaries and all sorts of organisations – for data collected from tracking and tracing activities constitutes a planetary ecosystem for governing people in their movements and activities. Whether this results in digital dictatorships or digital democracies arguably depends on a form of power that now conjoins other forms of power. Three technologies of power specially made sensory power sensible and intelligible during the pandemic: clusters, dashboards and tracking.

We have seen references to hotspots, epicentres and bubbles, all related to the spread of the coronavirus pandemic. These words indicate how infected or healthy bodies are related to each other. What makes these words performative is that the government's interest is in how these bodies are doing, not in punishing or disciplining them. Instead, they want to encourage them to achieve desirable outcomes. Governing performatives were invented for tracking these clusters to make rapid-response interventions. Consider the difference between governing gatherings like raves and parties, which elicited new policing concerns about discipline and punishment, and governing clusters, which are relations between bodies, not physical gatherings. While contact tracing programmes may not be entirely digital yet, tracking, testing and tracing bodies require frequent gathering, storage and transmission of data by various agents and authorities. Along with lockdown, distancing and isolation, tracking infected subjects, tracing those who may have had contact with them, and alerting clusters to isolate themselves, requires technologies of power appropriate to these objectives.

Where did these knowledges come from? Throughout the twenty-first century, states and corporations developed technologies to govern users of apps, devices and platforms. Data collection, storage and analysis from digital users led to competitive struggles between states, organisations and corporations. This competition also arose between professions like epidemiologists, statisticians, data scientists, programmers and security experts, who are transnational and whose expertise spans national borders.[7] Sensory assemblages, though not entirely digital, involve various digital technologies and analytics. For instance, mobility reports by Apple, Google and Facebook accumulate data about infections and deaths, which inform their tracking and tracing app toolkit. This demonstrates that tracking and tracing by corporations extend beyond disease infections to encompass activities like watching, listening, reading and communicating. While governing clusters to maintain health and wealth has proven elusive, coronavirus apps reveal a form of power that governs people through clusters.

Every form of power produces its visualisation, from cartography to anatomical diagrams and statistical charts. However, sensory power's visualisations have created novel technologies of representation. These visualisations identify unseen patterns and include interactive elements in dashboards that show the effects of combining data on cluster features. Dashboards have become a primary government technology, like cartography, anatomy and charts. Initially, representations like those on the Johns Hopkins University dashboard provided basic data and statistics mapped onto national borders. As the pandemic progressed, more sophisticated dashboards emerged for financial systems, transportation, military operations and managing football games. Rapidly developing dashboards are also in fields like migration and policing, used for governing movement and criminality. While dashboards on the pandemic were basic, they quickly became technologies of power funded by governments, developed by corporations and universities, evolving into more sophisticated forms. This brings us to the key difference between clusters: they are not continuous and instantaneous but live.

Sensory power organises algorithms for live measurement, identification, action and intervention. Data mobilise with immediacy and varying intensities and temporalities. It modulates body and people's performance through power technologies like algorithms. Instead of periodic statistics, people become live clusters with pulses, flows and patterns. Data serve three functions: identifying cluster features (e.g. infected, contacted, protected, vaccinated); monitoring and evaluating those features live (e.g. daily changes in R metric hotspots, epicentres, bubbles); and provoking changes in conduct (cajoling) or using sovereign or disciplinary interventions (easing or increasing lockdown). Identifying features is similar to classical data regimes: periodic measurement with indices, rates, metrics and indicators. However, governing clusters means live governing their pulses and signals.

This logic can be illustrated by tracking and tracing the virus's reproduction through apps and designing interventions like immunity or vaccine passports to govern clusters. Though these apps have failed, their logic will persist in other fields. The app tracks infected bodies, notifies, tests and isolates them to stop reproduction. It also traces bodies that encountered infected ones, notifying, testing and isolating them to slow reproduction. This creates a live cluster of infected or potentially infected bodies. However, governing people in clusters requires effective interventions at notifying, testing and isolating stages. This is a relation between regulative and disciplinary power: disciplinary technologies like consent for notification and self-isolation achieve the desired infection rate R. This is a costly and inefficient exercise of power. Yet the enthusiasm for a potential app and its promise to deliver a game-changer is palpable: to minimise disciplinary power and maximise sensory power. In other words, to formulate the problem of government as a relation between regulative and sensory power.

We have noted various stages of the cycle: tracking, notifying, testing, isolating, tracing, notifying, testing and slowing. If solutions automate testing and isolating, clusters could govern themselves. Technological limits exist, as do severe legal, political and cultural limits. However, these limits may be surmountable during the coronavirus pandemic or soon in other fields where they seem less relevant, like finance or migration.

We could not imagine such scenarios without technologies of power for tracking and tracing being already used in fields other than health. For forty years the accumulation of capital in finance, manufacturing, retail, transportation, hospitality, entertainment and other industries has been accompanied by the accumulation of peoples through tracking and tracing their movements and the modulation of sentiments, needs and desires. Similarly, fields of government such as policing, crime, migration, borders and education already feature technologies of sensory power. The live data produced from sensory assemblages pervade these sectors and fields. What we witnessed during the coronavirus pandemic is the acceleration of the development and deployment of technologies of sensory power in these fields over the last forty years.

Every form of power almost always betrays its governing intentions. For sensory power, code errs, algorithms misfire, data fail, apps malfunction. Yet each form of power produces affects and effects on people and elicits forms of resistance: sovereign power elicits revolt (protest, uprising, occupation), disciplinary power elicits subversion (strike, refusal, slowdown, anonymity) and regulative power elicits evasion (escape, deception). To put it emphatically, it is these struggles that make all forms of power sensible and intelligible. Each form of power draws forth what is a latent or potential resistance into a sensible and intelligible form. This is where the resistances against lockdowns, distancing, covering, isolation or vaccination misunderstood the relationship between sovereign power and regulative power: these technologies of regulative power proved effective in protecting the health of people in past epidemics and pandemics. Ostensibly resisting sovereign power without understanding its relation to regulative power proved a misguided and dangerous play with power.

What then are the forms of resistance that sensory power elicits? If indeed sensory power governs people through clusters, a problem of power becomes not only how to perform acts of revolt, subversion and evasion but to perform acts of resistance appropriate to sensory power in relation to other forms of power. We have so far witnessed concealment (encryption, anonymisation, aliases), opacity (spoofing, cloaking), and sentiments (irony) as forms of resistance. The remains of the pandemic demonstrate that the relations between states and citizens entered new arrangements in the twenty-first century with the birth of sensory power. The relations between and among the three forms of power have been transformed significantly with the birth of a fourth form. The relations between states and citizens are now mediated by a form of power whose logic permeates and penetrates states' territories, peoples and wealth from within and without. This changes not only the relations between the three forms of power but also their logics. We do not yet understand how these confrontations with sensory power relate to sovereign, disciplinary and regulative forms of power. How we can play with sensory power in relation to other forms of power is a question that this book articulates.

The games we play: kingship, citizenship, population, imitation

By taking a long view the book is organised around three games humans invented to displace the kingship game that emerged around 6,000 years ago with the first cities and states. The citizenship game emerged around 2,500 years ago with ancient polities with the invention of the autonomous citizen-subject. This invention is immensely complex and there have been numerous twists and turns to it for over a thousand years and beyond. But in essence the birth of the subject-citizen already embodies and resignifies sovereign and disciplinary forms of power of the kingship game. The population game proceeds in two stages. With the reorganisation of the kingship game about 500 years ago, the European empires assemble territories, wealth and populations as elements and constitute these three as objects of government. In the second stage, from about 250 years ago, European states invented a relationship between citizen-subjects and populations through regulative power. While certain elements of the imitation game can be traced to the second stage of the population game, it was invented only about 100 years ago by an imaginary of the human brain as a machine that could be produced.

The sources for this organisation are the style of thought on power we mentioned earlier, which insists on understanding organised polities not as a continuous history but as a fragmented yet persistent one of the last 6,000 years that is continually pushed back further with new works.[8]

James Scott, for example, argues that the emergence of sedentary polities (cities, states, nations and empires) should be seen as relatively recent developments against the long history of domestication of humans by humans (Baker 2019). For Scott the question is how humans 'came to live amid the unprecedented concentrations of domesticated plants, animals, and people that characterize states' (Scott 2017, 5). Scott has generated a considerable debate in history, archaeology and anthropology. An earlier generation of scholars such as Max Weber (1921) and Lewis Mumford (1961) considered cities, states and empires as distinct polities as 'civilisations'. The city, like other polities such as states, empires or even nations, is elusive to define. The city signifies a settlement, a settled people, a government or an imaginary space. This multiplicity also overlaps with other polities to make it difficult to signify the city unambiguously. It is possible to recast Scott by suggesting that the city is a way of organising social and political domination through which humans developed complex mental and physical capacities.

If indeed the city has been the embodiment of technologies of domestication of plants, animals and humans, it is also seen as a polity for the development of the human brain. This is a stronger hypothesis than that which Georg Simmel proposed on the relationship between metropolis and mental life (Simmel [1903] 1971).[9] Simmel was prescient in seeing the twentieth-century metropolis not only as a polity but also as a mental life. Simmel wrote on the cusp of the invention of the imitation game and before the development of neurosciences, and he did imagine how mental life developed through metropolitan life by focusing on aversion, indifference, empathy and anomie as distinct mental states in the metropolis.

This book revisits the historical and theoretical background against which the convergence between neurosciences and datasciences is creating a new polity and the citizen in the twenty-first century. We have previously described this historical and theoretical background as the city as a difference machine where we traced genealogically the development of sovereign, disciplinary and regulative forms of power producing both the city and the citizen in Europe and America (Isin 2002). We now revisit this hypothesis to consider if, and to what extent, neurosciences and datasciences afford us a concept with which to grasp how sensory power is shaping the polity and citizen in the twenty-first century with the invention of arguably an entirely new game: the imitation game.

If indeed datasciences in the twenty-first century increasingly imagine polities like nervous systems organised as neural networks with the development of artificial intelligence and algorithmic government, it becomes necessary to study the convergence between neurosciences and datasciences in the development of the twenty-first-century polity against this long view.

For datasciences, the polity in the twenty-first century increasingly appears as one that manages and governs itself autonomously (Zanella et al. 2014). There are numerous initiatives for creating sensory cities or transforming existing cities into sensory cities (Karvonen, Cugurullo, and Caprotti 2019a). These developments involve creating sensory ecosystems for managing and regulating transportation, housing and service infrastructures and monitoring, auditing, and controlling protocols (Lee, Hancock, and Hu 2014; Marsal-Llacuna and Segal 2016). With various slogans, these developments are converging where the imaginary city autonomously governs itself.[10] For some the autonomous city is a monstrous development and signifies the end of the city (Cugurullo 2021). For others, the smart city signifies a neoliberal logic for regulating and governing the citizen (Grossi and Pianezzi 2017; Cardullo and Kitchin 2019).

The approach of neurosciences to the polity has been different from datasciences but there are convergences. For example, the emergence of neurourbanism, or at least its propositions, imagine the city for the well-being of humans (Adli et al. 2017). Arguing that the city has always been associated with stress, anxiety and mental disorders, some suggest that 'a new field of neurourbanism addresses these issues, applying neuroscience laboratory methods to tackle global urban problems and promote happier and healthier cities' (Pykett, Osborne, and Resch 2020, 1). Yet in these initial stages neurourbanism focuses on addressing stress and anxiety that happens *in* the city so that the brain and the city are assumed to have independent but related lives. This rigid separation overlooks how the brain and the city are entangled developments. In recent approaches to neurosciences and architecture (or neuroarchitecture) there is an emerging recognition of this entangled development (Eberhard 2009; Myers 2020).

While neurourbanism or neuroarchitecture have raised questions about the ways in which datasciences create an imaginary of the city as an autonomous polity, the dominant tendency is still to envision the brain receiving and the city generating and transmitting various stimuli. The imaginary of the smart city has been instituted by data collected by sensors throughout the city for specific monitoring and controlling systems. Yet the development of mobile crowdsensing technologies is increasingly incorporating data collected from mobile and immobile sensors embedded in ubiquitous devices.[11] The consequences of this development are yet dimly visible. But we know how data collected from the activities and movements of mobile people augmented by data from immobile sensors are transforming cities and citizens simultaneously.[12] Although a failed project, a proposal by Google to build a city on Toronto's waterfront reveals this convergence between immobile and mobile sensors in tracking, tracing and documenting the activities and movements to create a city as an autonomous polity governing itself.[13] The Sidewalk development in Toronto signals the entry of platform and app developers with crowdsensing technologies into making and remaking cities. As some scholars have recently suggested, the relationship between metropolis and mental life that Simmel proposed in the early twentieth century takes on a transformative turn in the early twenty-first century.[14]

There is no doubt that the city as we know it is being radically transformed and the imaginary of the smart city is related to the production of a new citizen-subject. This book is diagnostic: the city and the citizen yet to come are open to interventions and transformations and the emerging imaginary of the city in datasciences and neurosciences provides a glimpse into how this edifice has been built. We will consider the city as a species of polity and explore its entanglements with empires, states and other polities. We use 'polity' (or polities) as it has been used in European political and legal discourse since the twelfth century: a multitude of people combining and integrating into a collective body with a generative will greater than the sum of the constitutive wills of the multitude. At the original scene of ancient and modern polities were political agents as autonomous subjects. At the original scene of learning machines are intelligent agents as autopoietic agents.

We will revisit this narrative arc in the conclusion once we describe the apparatus that sensory power created as meticulously as possible. How forms of knowledge stimulate, mobilise and modulate laboratory and experimental practices, create concepts, organise practices, invent methods and generate objects and subjects in, of and by doing things with sensory power is the subject of this book. We will examine how sensory power generates two forms of knowledge (neurosciences and datasciences) that assemble new kinds of polities. However, as we shall see, sensory power is nested within and operates alongside earlier forms of power (sovereign, disciplinary, regulative), resignifying, regenerating and recharging them in ways that also transform their meaning and function, and reorganising earlier forms of knowledge associated with them. Before we can understand ways of doing things with numbers and neurons, we need to outline analytics of doing things with power.

Notes

1. Both as a noun and as an adjective, the word 'game' is one of the oldest and most diverse words in the English language. The Oxford English Dictionary records forty-nine distinct meanings since at least the twelfth century. Originally, the word 'game' referred to an activity that provided amusement, diversion, or pastime, such as playing athletic sports or hunting. In everyday language it gradually signalled playing games as both serious and playful contests that require reasons, emotions and some mastery over its norms. Not surprisingly, game theory, which originated from computing, treats playing games as a rational (propositional) rather than affective and effective (performative) activity. The path we have followed was opened by Alan Turing's imitation game, which introduced playfulness; J. L. Austin's (and Ludwig Wittgenstein's) language games, which introduced undecidability; and Pierre Bourdieu's games, which introduced

power. As we will see, game theory was never a sideshow for but central to transforming Alan Turing's imitation game into a rational game.

2. Watson and Breedlove's *The Mind's Machine* ([2012] 2021) is a believer; Boden's *Mind as Machine* (2006) is sceptical; Dreyfus and Dreyfus's *Mind over Machine* (1986) is critical. All nonetheless assume the mind and the machine have had separate if independent histories – and assume that we are critical.
3. Ian Hacking (2002) named this style of thought as a playful irony. He was inspired by the concept of genealogy invented by Nietzsche ([1887] 2007), articulated by Deleuze ([1962] 1985) and practised by Foucault ([1971] 1977). This is because ontology is meant to be a study of timeless and spaceless being while history is unthinkable without times and spaces of beings. Yet this playful irony performs genealogy as a strategy.
4. We are drawing from Isin and Ruppert (2020a, 45–57).
5. J. L. Austin's influence on contemporary political theory has been significant (Bourdieu 1991; Butler [1997] 2021; Derrida [1977] 1988; Latour 2000; Felman [1980] 2003). Austin was also a lesser-known influence on Foucault and Deleuze.
6. We are drawing from Isin and Ruppert (2020b). With our experience of the pandemic, this is where we first named this form of power with our experience of the pandemic.
7. Francisca Grommé, Evelyn Ruppert and Baki Cakici discuss (2018) data scientists as an emerging class of analysts.
8. It is remarkable to compare just the first two volumes of the *Cambridge World History* with twentieth-century scholarship on world civilisations (Barker and Goucher 2015; Christian 2015).
9. See the discussion by Nikolas Rose and Des Fitzgerald (2022) of Georg Simmel and the twentieth-century sociology between mental life and the city.
10. A critical literature is growing rapidly on smart cities. We will discuss it in more detail in Chapter 4. For now see Albino, Berardi and Dangelico (2015); Kummitha and Crutzen (2017); Angelidou (2015).
11. See Ganti, Ye, and Lei (2011); Fiandrino et al. (2017); Capponi et al. (2019).
12. See Cardone et al. (2014); Kong, Wang, and Chen (2019); Leppänen et al. (2017); Yang et al. (2016).
13. We will see in Chapter 4 that the literature on the imaginary of the smart city is large and concerns cities in all continents of the planet earth. For Toronto see Spicer and Zwick (2021).
14. See Fitzgerald, Rose, and Singh (2016); Rose and Fitzgerald (2022).

1

How to do things with power: words, numbers, neurons

Naming the birth of a new power – sensory power – requires reworking analytics of power that Spinoza and Nietzsche initiated and Foucault and Deleuze elaborated. We begin with three forms of power (sovereign, disciplinary, regulative) and associate these forms with forms of knowledge (datasciences, neurosciences) with which power creates affects and effects, and forms of capital (cultural, economic, symbolic) that enable them. This is immensely difficult not only because both Foucault and Deleuze were drawn to other projects but also because sensory power was barely sensible or intelligible in the 1970s and 1980s, let alone disciplinary power in the 1670s (Spinoza) or regulative power in the 1870s (Nietzsche). We need to rework analytics of power rigorous enough to identify distinct elements of sensory power while remaining truthful to the project against a flat concept of power that (a) considers it a property that can be possessed and (b) is inattentive to its separate forms and how these forms dynamically change over time in relation to each other. This is why the project of 'doing things with power' is not about proposing a theory of power but separating power as such from its forms and investigating these forms as performatives rather than properties.

We will call the former power (in) theory and the latter power (in) practice. This will help us rework analytics of power, as we are convinced that the inattention to separating power as such from power as performative is the reason for flattening of the concept of power. This inattention arguably also led Spinoza, Nietzsche, Deleuze and Foucault to make conflicting claims. In many ways all four can be considered reluctant theorisers of power in the sense that none actually set out to theorise power as such but wanted to understand how people perform power. Our reworking begins with Austin by focusing on doing things with power and attempting to capture this approach to power not as a substance or thing but as something that is brought into being by exercising or performing it.

If we are equally reluctant to theorise power as such but nonetheless recognise that articulating a concept of power in order to examine its performatives cannot be avoided, we need an analytical approach to resolving this problem. Developing the concept of 'doing things with power' requires articulating both power as such and power performatives (ways in which it is exercised). This claim can only be evaluated by trying or using it, without promises – just as the most effective understanding of a word is not merely or only its meaning but also its function or how it is used. To separate two different ways of talking about power, we will acknowledge that power as such does not exist, yet we can talk about it, and that talking about power as performatives means analysing how power is exercised, and the forms, strategies and technologies by which 'it' exercises affects and effects.[1]

All this is further complicated by the fact that analytics of power is not just about doing things with power, but there are different things like words, numbers and neurons that power does things with. We are more familiar with doing things with words that Austin (1962) invented to describe how in, of and by doing things words make things happen. Yet Austin could not possibly consider doing things with numbers or doing things with neurons. His work leaves open the possibility that words never meant narrowly spoken words – speech acts – but also included gestures, movements and many other things that convey meaning through their function. We are now attempting to think about doing things with numbers and doing things with neurons as we make our claim that a new form of power has been invented in, of and by doing things with numbers and neurons and that it poses the greatest challenge yet to political thought.

Power (in) theory

The speaking of the word power immediately evokes and provokes affinitive or associative words: authority, command, domination, legality, legitimacy, obedience, resistance, submission, subversion and so on. It also evokes words about forms of government through which power performs functions: authoritarian, autocratic, democratic, despotic, oligarchic, tyrannical and so on. It is impossible to speak about power without (intentionally or unintentionally) mobilising some or all these concepts as synonyms, antonyms, metaphors or allegories of power. Yet power is irreducible to its affinitive and associative words or forms, strategies and technologies with which it is exercised or affects and effects that its exercise produces.

This perhaps attests to the fact that when we speak about power we are speaking about how beings (and things) arrange and organise their existence together in, of and by doing things with power. If power is the name for how beings and things arrange and organise their existence of being thrown together then it is utterly ineffective to ask 'what *is* power?' without examining what is naming power and for what purposes, for power is named and exercised simultaneously. Conversely, we cannot even begin with power as such or in general but how it activates beings and things to arrange and organise their existence together in, of and by doing things with power. Foucault often insisted that power in general (or as such) did not exist, and that how power is exercised must be the object of research, and yet he also spoke often enough about power in general, sometimes mentioning affects and effects of power and sometimes its features in general. Foucault insisted that power is a relation, that it is not a property of any agent, that it is an action upon action, and that its aim is to effect conduct.[2] These are of course ways of speaking about power as such to prepare the ground for analysing how all this happens in practice when it is exercised and named – in other words, performed through different forms. If Foucault investigated how power was exercised, naming different forms of power, identifying their mechanisms and apparatuses, periodising their beginnings, transformations and articulations, and identifying their affects and effects, especially forms of government, forms of knowledge and forms of subjects, he was also inventing a way to speak about power as such. Over twenty-five years, Foucault continuously worked and reworked different ways with which to speak about power as such and power as performatives as his body of work approached its relations to knowledge and subjectivity, which he always argued that were essential to his analyses, from different angles and with different emphases. We often find Foucault speaking of power in relation to either knowledge or subjectivity as he insisted that there is a generative tension between forms of power and forms of knowledge and forms of power and forms of subjectivity.[3]

If we are speaking about the birth of a new form of power – sensory power – we need to untangle these different ways of speaking about power, its forms, its affects and effects, the strategies and technologies with which it is exercised, and the activities of the government of people that it produces. We will consider 'power as such' as a way of speaking about power that makes identifying its general properties possible and 'power as performatives' as a way of speaking about power when it is exercised through strategies and technologies, which produces affects and effects.[4]

The phrase we used above – 'how beings and things arrange and organise their existence together in, of and by doing things with power' – now

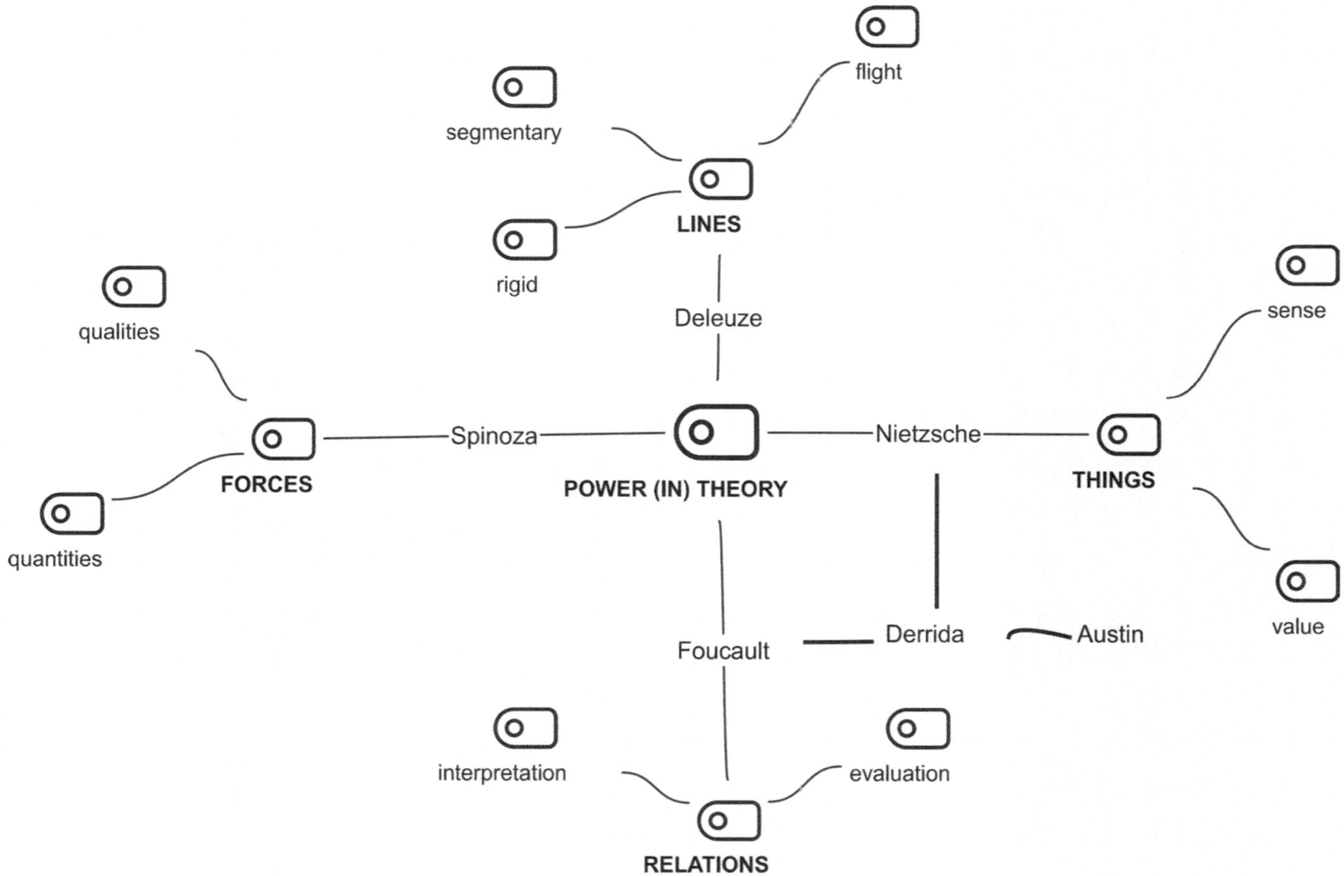

Figure 1.1 Power (in) theory or will-to-power

requires elaboration as a key analytical approach of this book.[5] Deleuze's Nietzschean concept of power as such (the will to power) will set the stage for interpreting and evaluating different forms of power, affects, effects, strategies and technologies as its performatives.

For Deleuze's Nietzsche, the starting point for the analysis of power is the body. To speak of the body is to speak of the forces that compose it. Bodies can be both natural (organisms) and artificial (machines) or singular (individuals) and collective (polities).[6] The determination of forces that endow bodies with their capacities and drives is the second aspect of the analysis of power. A body's ability to affect or effect another body depends on the forces that compose it, which manifest themselves as capacities or drives. We must therefore understand what forces compose bodies that effect or affect each other.

To understand the composition of forces and how they endow the body with capacities and drives, we can measure them by the quantities and qualities of these forces. The quantities range across weak and strong spectrum in relation to forces on which the body exercises its capacities and drives. We can also measure qualities of the forces across active and reactive range. Active qualities of forces are affirmative: plastic, irrepressible and differential. By contrast, reactive forces are negative: utilitarian, incapacitant, autoimmune. The forces can be dominant or dominated in relations to other forces, which will determine the capacity or drive of bodies that they compose but also the direction of their strength whether active or reactive.[7] Obeying and commanding are forms of struggle that the quantities and qualities of forces will shape and whether they will be dominant or dominated.[8]

As Deleuze says, Nietzsche (see *On the Genealogy of Morality*, [1887] 2007) provides examples from across his work in the 1880s to illustrate what he means by force, how a composition of forces endows a body with capacities and drives, how these forces can be weak or strong, active or reactive, how a composition can be dominated or dominating in relation to other forces, and how commanding or obedience are effects of these relations. Nietzsche does not initially interpret or evaluate these forces. Whether a body composed of forces is weak or strong or active or reactive in relation to another body are not performative (interpretive or evaluative) statements but are constative statements. They become performative utterances when we begin to interpret and evaluate the affects and effects of these relations.

This requires making judgements about the quantities and qualities of forces that compose bodies, endowing them with capacities and drives to act upon each other. These judgements will arise from struggles between forces for domination or emancipation. For Deleuze, speaking about power as such (the will to power), Nietzsche was introducing sense and value into thinking about power. Deleuze would say that 'the sense of something is its relation to the force which takes possession of it, the value of something is the hierarchy of forces which are expressed in it as a complex phenomenon' (Deleuze [1962] 1985, 8). We thus enter into two activities that mobilise any struggle between and among bodies and the forces that compose them: interpretation and evaluation. The activity of interpretation produces the sense of the body in relation to the force acting upon it and the activity of evaluation determines the differentiation between forces that compose it. These two activities generate power as such (the will to power) as the capacity and drive of forces that take possession of a body and generate a difference between forces that compose it.

As Deleuze says, 'for all these reasons Nietzsche can say that the will to power is not only the one that interprets but the one that evaluates' (Deleuze [1962] 1985, 54). This is because 'to interpret is to determine the force which gives sense to a thing. To evaluate is to determine the will to power which gives value to a thing' (54). Thus, 'the will to power as genealogical element is that from which senses derive their significance and values their value' (54).

The concept of the will to power would prove most challenging after Nietzsche created it. For Deleuze, '*The will to power is the element from which derive both the quantitative difference of related forces and the quality that devolves into each force in this relation.* The will to power here reveals its nature as the principle of the synthesis of forces.'[9] By rephrasing this as 'force is what can, will to power is what wills' (50), Deleuze introduces Nietzsche's concept of genealogy, in which force is analysed by its qualities and quantities and power is analysed by its interpretive and evaluative capacities.

Deleuze would understand genealogy as a style of thought that studies differential and genetic elements of forces that compose the body and its relations to other bodies. This is because 'the will to power is the differential element of forces, that is to say the element that produces the differences in quantity between two or more forces whose relation is presupposed' (52). And 'the will to power is the genetic element of force, that is to say the element that produces the quality due to each force in this relation' (53). The forces will be dominant or dominated according to their difference in quantity, measured as weak or strong, and they will be active or reactive according to their quality.

If indeed 'to interpret is to determine the force which gives sense to a thing [then] to evaluate is to determine the will to power which gives value to a thing' (54). We can thus understand 'the will to power as genealogical element is that from which senses derive their significance and values their value' (54).

If genealogy is a style of critical thought, it accepts that interpretation and evaluation are challenging activities. We must judge which forces are superior or inferior, active or reactive, and whether they will perform as dominant or dominated (58).[10] We must recognise that 'the will to power does not mean that the will wants power. Will to power does not imply any anthropomorphism in its origin, signification or essence' (85).[11] This is why 'the will to power must be interpreted in a completely different way: power is the one that wills in the will. Power is the genetic and differential element in the will' (85).

A question arises here with respect to interpretation and evaluation: we can understand how active forces become dominant, but can reactive forces become dominant? 'When they get the better of active forces do reactive forces themselves also become dominant, aggressive and subjugating? Do they, by getting together, form a greater force that would then be active?' (57). Deleuze says that: 'Nietzsche's answer is that even by getting together reactive forces do not form a greater force, one that would be active. They proceed in an entirely different way – they decompose; they separate active force from what it can do; they take away a part or almost all of its power' (57).

And it is here that Deleuze introduces a notion that he will repeat for years to come, which is an analytical approach but it is also an ethical and political position. This notion comes from Spinoza and Nietzsche: that reactive forces become dominant in an entirely different way by decomposing active forces and separating them from what they can do. For Deleuze, 'an active force *becomes reactive* (in a new sense) when reactive forces (in the first sense) separate it from what it can do' (57). This is an interpretive and evaluative statement that becomes a performative utterance. We must always interpret it by showing how power is exercised in its affects and effects. Deleuze would recognise that 'it is primarily in this sense that interpretation is such a difficult art – we must judge whether the forces which prevail are inferior or superior, reactive or active, whether they prevail as dominated or dominant. In this area there are no facts [statements], only interpretations [performatives]' (58).

If Nietzsche claims that power as such (the will to power) is a new way speaking about power (Deleuze [1962] 1985, 80), what are its consequences? First, it transforms the speaking subject from an observer or spectator into an interpreter and evaluator. Speaking about power as such means that power is not an object of representation but activation of forces that compose it (80). Second, power as such is not interpreted or evaluated according to an external factor: 'when we make power an object of representation, we necessarily make it dependent upon the factor according to which a thing is represented or not, recognised or not' (81). Third, power as such cannot be explained by power as performatives such as struggle, war, combat and battle; it is power as such that interprets and evaluates these performatives (82). As Foucault will memorably say later, these principles implicate power in all forms of knowledge.[12]

To put it differently, speaking about power as such and separating it from power as performatives establishes a different relationship between power as such and the speaking subject. First, since speaking the truth is interpretation and evaluation, truth is generative and performative (Deleuze [1962] 1985, 103). Second, the speaking subject has both affective (passions, desires, pleasures) and effective (evidence, signs, facts) relations to truth (103). Third, arriving at truth as interpretation and evaluation cannot be a guaranteed course of action or method. To put it yet more succinctly, truth is not an abstract universal that can be repeated but a force in the formation of power as such (the will to power).

If genealogy is a way of speaking about power as such as its genetic element, we must devise ways of analysing how power is exercised as its differential element. This is the intervention of Foucault. First, he identifies affects and effects of power as performatives. Foucault analyses power as performatives in how power produced such effects as not only coercion, repression, oppression, domination and freedom and resistance but also how it produced such affects as not only terror, repulsion and resentment but also pleasure, joy and desire. Second, Foucault meticulously documents strategies of power such as normalisation, direction and control and technologies of power such as spectacle, examination, inquiry, survey, classification, sequestration and detention. Third, Foucault, like a cartographer, as Deleuze ([1986] 1988) would say, maps these affects, affects, strategies and technologies of power to several forms that he signifies such as sovereign, disciplinary and regulative power, thus both periodising and spatialising them. Foucault demonstrates how reactive forces become dominant in an entirely different way by decomposing active forces and separating them from what they can do through sovereign, disciplinary and regulative forms of power.

For Foucault, sovereign power had its origins in the beginnings of polities about 6,000 years ago but it became particularly pervasive 500 years ago (Foucault [1973] 2000, 30). Similarly, disciplinary power could be found in earlier polities, especially 2,500 years ago, but it became pervasive about 250 years ago (Foucault [1973] 2000, 30–2). For Foucault, periodising and spatialising regulative power is most challenging. When he names disciplinary power, he speaks about it in both active and reactive qualities but, as his work develops, he perhaps makes the most influential discovery when he names it first regulative and then biopower and then government – all of which open various aspects of this form of power to analysis. We shall discuss these three forms in more detail below but we need to explain how the introduction of the concept of government into analytics of power was a crucial invention.

If indeed the will to power does not imply any anthropomorphism in its origin, then the question arises about naming a form of power that especially concerns the conduct of people. That concept will be government. As Foucault says, while for most of its history 'government' in ordinary and official languages meant, for example, governing children, women, patients and soldiers, it was only about 500 years ago that government became associated specifically with the state as a polity. Using 'government' in a broader historical sense, Foucault insists that overall 'one never governs a state, a territory, or a [polity]. Those whom one governs are people, individuals, or groups' ([2004] 2007, 122). We wonder if introducing government was (consciously or unconsciously) a way for Foucault to separate power as such from power as performatives. For once we begin speaking about power as performatives, we shift our emphasis from beings and things as such to peoples. At any rate, this introduces an additional dimension into genealogy of power as we have seen as it now involves governing people. We will use power when it involves both peoples and things and government specifically when we are speaking about exercising sovereign, disciplinary, regulative and sensory power by people over people.[13] This will enable us to avoid the assumption that power relations are always between two or more human bodies – anthropomorphism. It is not helpful to analyse power as such as though it is only between two or more bodies involving intentions and purposes. As we have seen, power relations are always between and among bodies, that bodies both natural and artificial can exercise power, and that bodies are composites of various forces: biological, neurological, chemical, economic, social, cultural, political etc. So, when we are examining power as performatives, we are examining power relations between and among bodies composed of forces. When these forces are activated by and over people, we can speak about governing peoples though not separated from power as such but as narrower organisation and arrangement of power as performatives among humans. As much as it is most helpful to think about government as an activity of governing people, and focus on power as performatives, we must still maintain the principles of analysing power as such. The challenge is to maintain a difference between power as such (the will to power) and power as performatives on the one hand and between power and government on the other hand. If we meet this challenge reasonably well throughout this book, we will have the opportunity to interpret and evaluate its effectiveness in giving an account of the birth of sensory power.

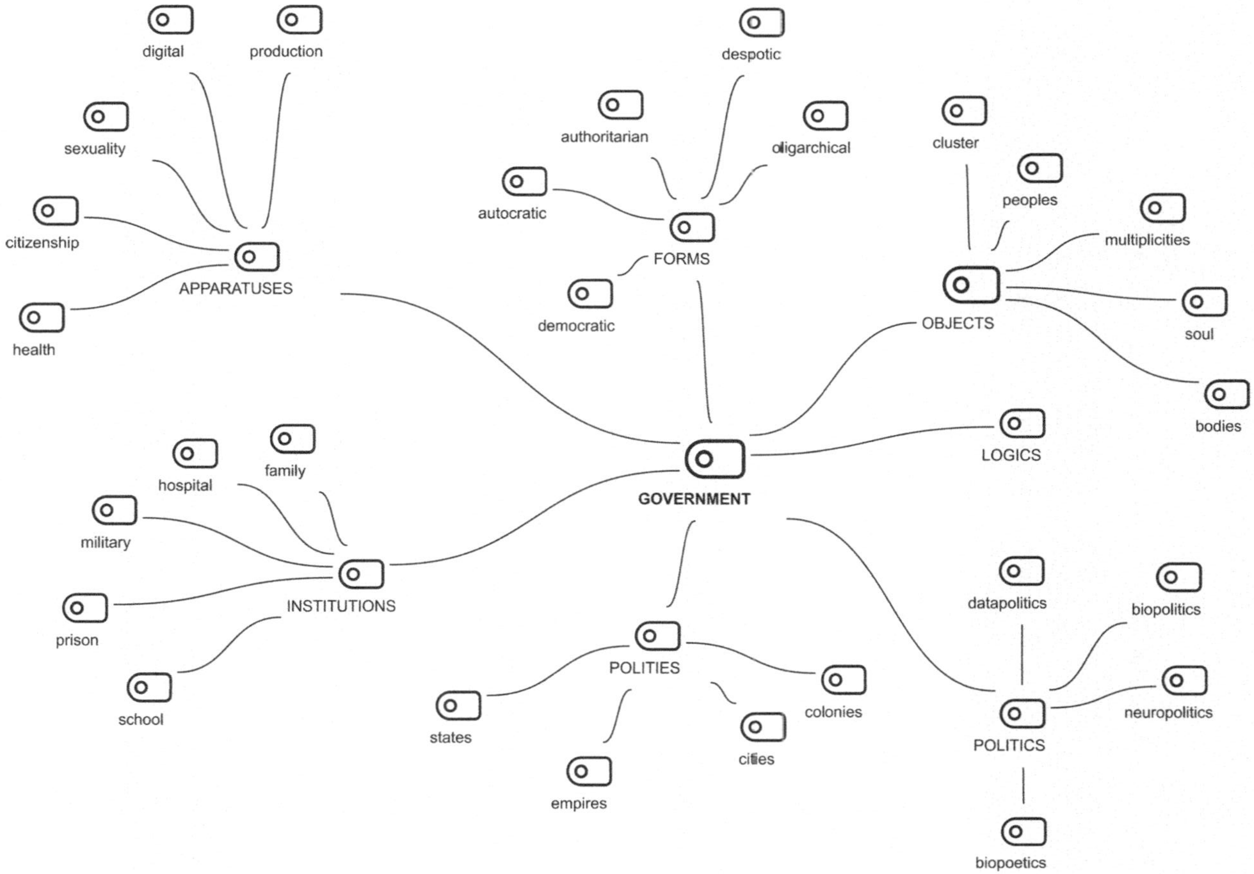

Figure 1.2 Government, logics, forms, institutions, polities

Foucault's most important discovery about government is to separate the logics (e.g. positive, negative, generative, prohibitive, repressive, liberating) and apparatuses (e.g. production, security, health, wealth, citizenship) of government from its forms (e.g. autocratic, democratic, despotic, theocratic) and institutions (e.g. family, hospital, military, prison, school). Foucault also carefully separates the objects of government (e.g. body, soul, people, population, species) from polities (e.g. cities, states, colonies, empires).

Taken together, analysing power as such (the will to power) and power as performatives, analysing government logics and apparatuses, and the forms, institutions, objects and polities that it produces are vital aspects of historical ontology or analysing the ways in which we have become what we are or analysing how people and things have arranged and organised their existence together in, of and by doing things with power.

Power (in) practice

The specific form of power – sensory power – emerged around 100 years ago and became pervasive in arranging and organising beings and things together in, of and by doing things with power about 40 years ago. It was in the 1930s that the drive for a computing machine for modelling it after an imaginary of how the brain works began bearing results.[14] These two drives organised two fields of knowledge: datasciences and neurosciences. We will develop a concept of sensory power being made possible by datasciences and neurosciences that governs people and things through continuous modulation of conduct between bodies and clusters – concepts, practices and methods that we shall elaborate in Chapters 2 and 3. Perhaps sensory power has become pervasive enough that it is shaping the polity and the citizen-subject in the twenty-first-century world. We have been experiencing this through our everyday lives and the pervasive discourse on this experience. Our readers will be familiar with artificial intelligence, machine learning, big data and their uses in politics, war, production, consumption and exchange. They will also be familiar with various debates on data colonialism, data inequality and data discrimination, as well as manifestations of these technologies in constructs such as smart homes, smart cities and smart worlds. Our argument is that the tectonic shift that makes all this possible is the birth of a new power – sensory power – that has a long history embedded in the history of other forms of power and power as such. How this new form of power has come to be, how it manifests itself in practice, and how we can trace it back to power (in) theory or power as such are questions that should enable us to explore a new perspective on what we have become all too familiar but are at the same time estranged from.

This means avoiding an assumption, and this is one of our challenges, that other forms of power – sovereign, disciplinary, regulative – have either disappeared or become less pervasive. Or that the games of kingship, citizenship, population and imitation replaced each other successively. On the contrary, all forms of power and games articulate together, their functions mutating and transforming in relation to each other. When we talk about power in a specific space and time, we must also specify how its multiple forms are articulated to arrange and organise collective existence. On the other hand, if we are to say anything about power as such (the will to power), we must strictly separate this from how power is exercised or performed in a space and time specific domain or field and its form – sovereign, disciplinary, regulative or sensory.

If we insist that we must study power by only how it manifests itself in a specific time and place, we will always be driven to speak about that which manifests itself and whether it has its own history. The tension between power as such (will to power) and power as performative will remain a tension to break through the past and present forms of power and their articulation. To put all this differently, that which we name sovereign power after disciplinary power was invented is not the same power which we name sovereign power. Similarly, we will need to recognise that what we name disciplinary power before regulative power was invented is not the same. If we are analysing power as such (the will to power) and power as performatives, analysing government logics and apparatuses, and the forms, institutions, objects and polities that it produces are vital aspects of historical ontology or analysing the ways in which we have become what we are or analysing how people and things have arranged and organised their existence together in, of and by doing things with power are constantly shifting fields of operation in relation to each other.

With these warnings, we can now provide an outline of different forms of power (sovereign, disciplinary and regulative) and name the birth of a new form of power – sensory power. Understanding this power and its relations to forms of knowledge that it both precipitates and is generated by provides the background to understanding the emergence of neurosciences and datasciences. We will first summarise how these three forms of power have accumulated over time and how, rather than supplanting each other, they articulated into one another. Then we will provide a preliminary outline of an argument that in the last forty years or so, a new form of power has emerged in connection with new forms of knowledge: neurosciences and datasciences.

This outline of the forms of power is necessary since their periodisation or operations remain rather imprecise. Foucault's studies in the 1970s

([1976] 1978; 1979; [1975] 1977), where he understood power as strategies and technologies through which people govern the conduct of others and selves, proved extraordinarily productive.[15] Foucault dramatically expanded the concept of government from states to various sites such as clinics, workhouses, hospitals, armies, prisons, camps, schools, cities, and spaces where governing the conduct of others and selves precipitated inventions of strategies and technologies by which it was accomplished (Lemke 2019). The overall picture is complex but nevertheless we attempt to encapsulate them to organise analytics of power. Power (in) practice names to capture what it organises, deploys strategies and technologies, produces knowledges, generates objects it governs, assembles apparatuses, and elicits resistances through which it governs.

When we say a form of power, we mean a logic through which bodies produce affects and effects by doing things with words, numbers and neurons. The three forms of power are identified by the name given to their distinct logics: sovereign, disciplinary and regulative. Like Foucault, we use strategies to indicate various actions that are purposive but non-subjective. There are strategies of power in and through which subjects purposively take positions but which are not a product of any singular or collective subject. Like Foucault, we refer to technologies in a broader sense to include not only devices, computers, services, switches and routers but also ways of organising actions and practices. As we have seen above, strategies and technologies give rise to specific knowledges necessary for their exercise and realisation such as those of scientific disciplines (e.g. medicine) and their objects of government (e.g. population). All forms of power are exercised through apparatuses such as health, wealth, security and citizenship, and assembled by relations between technologies, concepts, practices, methods, agents, authorities, institutions and so on. And finally, while people are subjects *to* power, each form elicits different *resistances* from people as subjects *of* power such as revolt and evasion.[16]

A key insight on these three forms of power is how the 'accumulation of capital' is intertwined with the 'accumulation of subject peoples'. These two phrases expressed in connection with forms of power may surprise some readers but it is Foucault ([1975] 1977, 220–1), who insists that 'the two processes – the accumulation of [subject peoples] and the accumulation of capital – cannot be separated; it would not have been possible to solve the problem of the accumulation of [subject peoples] without the growth of an apparatus of production capable of both sustaining them and using them; conversely, the techniques that made the cumulative multiplicity of [peoples] useful accelerated the accumulation of capital'. The accumulation of subject

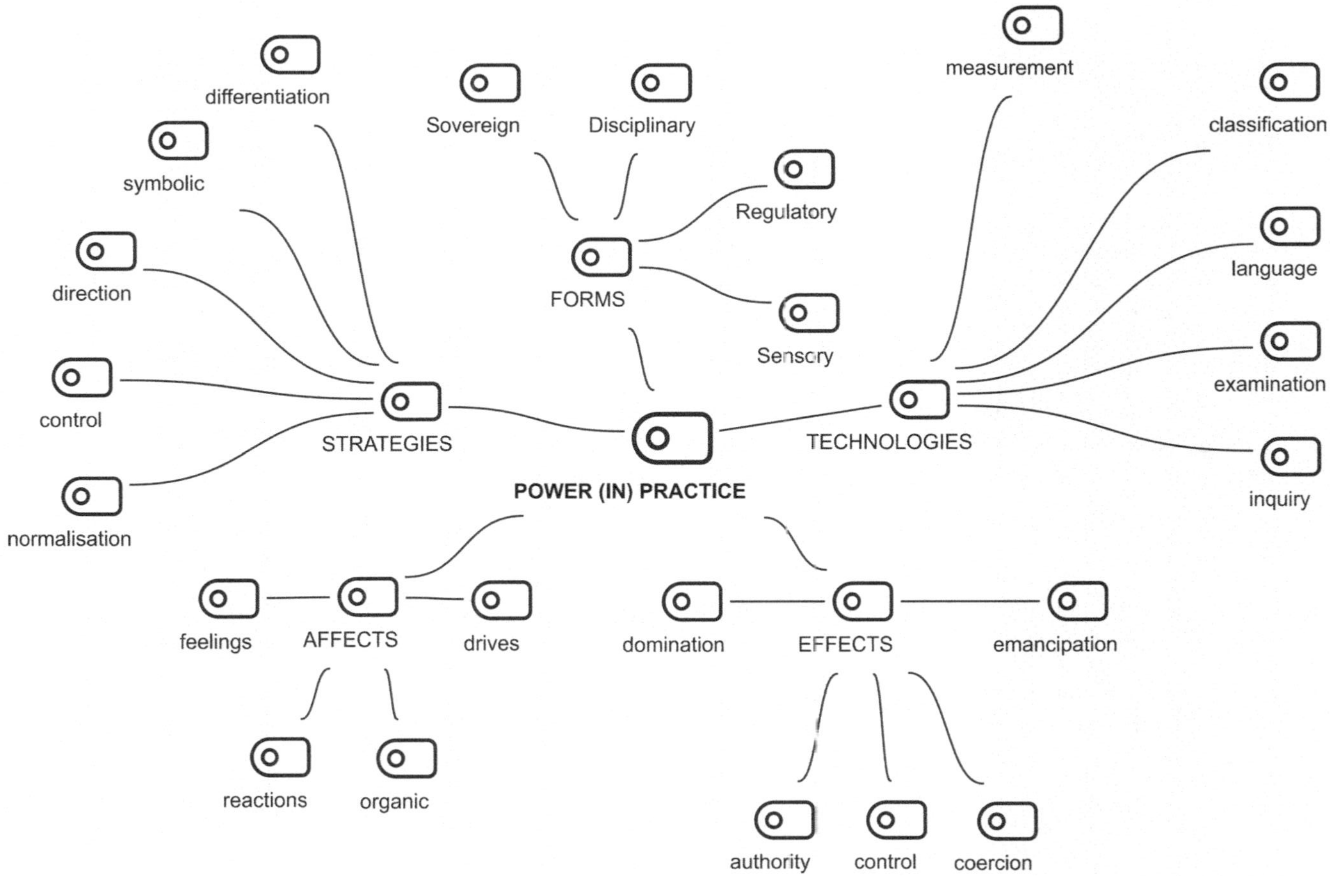

Figure 1.3 Power (in) practice or power performatives

peoples (making multiplicities of peoples useful, healthy and productive) and the accumulation of capital (generating economic, cultural and social capital and transforming it into wealth) also require the accumulation of knowledge (about objects and subjects of power) appropriate to these forms. These relations require, as Foucault ([1977] 1980) expresses many times, that power as performatives functions both negatively (cruelty, threat, fear, dread, torture, despair) and positively (pleasure, attraction, seduction, fulfilment, hope).

Foucault ([1975] 1977) insists that sovereign forms of power were always ritual, costly and violent. We take him to mean sovereign power before disciplinary power was invented. This is because sovereign power can be a more pervasive form than one that is ritual, costly and violent. After the invention of disciplinary power its modes of extraction of obedience are not necessarily or always rely on costly and violent forms. There are always various negative and positive repertoires available for its exercise and sovereign power itself has gone through changes over time (a point we develop later).

Foucault also expresses that sovereign powers 'soon fell into disuse and were superseded by a subtle, calculated technology of subjection' ([1975] 1977, 220). Gilles Deleuze ([1990] 1995a; [1990] 1995b) generalises this expression by arguing that 'disciplinary societies' were giving way to 'control societies'. These claims of supersession partly arise from conflating power as such and power as performatives. We must avoid thinking that forms of power fall into disuse and are superseded by new forms. We would rather interpret and evaluate how new forms of power articulate into existing forms, nestle within them for periods, and often mutate into new forms. Foucault ([1976] 1978) sometimes expresses this as a passage from one form of power to another with overlaps but the assumption of supersession came up often enough. As James Scott (2017) suggests, all forms of power may have existed in incipient and nestled forms since the origins of cities, states and empires as organised polities. There is a limit to focusing only on power that emerged about 500 years ago in Europe while longer genealogies of power provide a more complex picture.

There is also an implicit but a key claim in Foucault's studies where modern forms of power, especially modes of subjectification, are symmetrical. This is partly borne out of Foucault's brilliant intervention that power functioned not only negatively but also positively, but which forced his thought to project an image where all bodies act on the actions of other bodies in a level playing game. This may have resulted from a challenge in separating power as such from power as performatives, but Foucault sometimes speaks about the latter as though it applies to the former. In Foucault's studies, asymmetries – which are themselves also both products and enablers of power relations, such as class, gender and race – play rather ambiguous roles (Stoler 2016a; 2016b). Lastly, and this is partly a consequence of focusing on power as such in Europe, Foucault limits himself to 'societies such as ours' while overlooking how 'societies such as ours' were fashioned through the domination, dispossession and oppression of 'other societies'. To put it differently, as Walter Mignolo (2011) illustrates, modernity and coloniality were two aspects of the same development in different forms of power and its neglect is an important limit in Foucault's studies on power. Empires experimented on subject peoples with various strategies and technologies of power in colonies that were taken up in the metropole. It is impossible to isolate 'societies such as ours' from 'societies such as theirs' and this recognition has immense consequences for studying power.

There is a complicated matter of how Foucault articulates, develops and broadens his studies on forms of power during his lectures that were published posthumously and whether these lectures provide elaborations beyond the published studies.[17] It is vitally important to recognise that the following summary is therefore *our* reworking of the three forms of power as we claim that the fourth form of power has become sensible and intelligible with the concept of doing things with power as the principle.

We have suggested above that the coronavirus pandemic illustrates how each form of power functions through several strategies and technologies, forms of knowledge and objects of government. There are both disadvantages and advantages of illustrating from a recent and pervasive event. The interpretation and evaluation of such events continue and change as both new evidence is brought to bear on them and new perspectives challenge them. This makes it difficult to draw any definitive conclusions. Yet there is something about both power as such (the will to power) and power as performatives as they develop over long periods and their practices become ritualised, routinised and regularised, that their effects and effects, their strategies and technologies may no longer appear as power relations. A disruptive event such as a pandemic reveals these operations as authorities become concerned about the consequences for the order of things. The pandemic, partly because of the intensity of the experience over two to three years and partly because of the geopolitical disruption, revealed various aspects of the three forms of power and indeed provided a glimpse of a fourth. Or, rather, how three forms of power had been imperceptibly transformed over the period when a fourth emerged. We have observed how sovereign power extracts obedience (lockdown, curfew, containment), disciplinary power demands submission (sacrifice, distancing, isolation, punishment, hygiene) and regulative power calibrates the effects of obedience and submission between the health of populations (infection rates, transmission speed, mortality rates, recovery pathways and immunity rates) and bodies. Each of these three forms of power can now be observed as a segue to our discussion of a fourth form of power that we think is entwined with but distinct from these three forms in the present.

How to do things with sovereign power

From 500 to 200 years ago, the rebirth of sovereign power is associated with the birth of modern imperial and state apparatuses by which they were governed. During this period, sovereign power became increasingly concerned with the health and wealth of subject peoples. European empires were built through the accumulation of subject peoples through slavery, the colonisation of indigenous peoples and the settlement of colonies. If each of these movements required the extraction of obedience from subject peoples, it also precipitated the search for more effective and efficient ways of governing them. If the key objects of government were territories governed through technologies of settlement, deportation and dispossession, key forms of knowledge also developed on the wealth and health of subject peoples known as political arithmetic (Petty [1682] 1888). The accumulation of mercantile capital would have been inconceivable without the transatlantic slave trade of African peoples into colonial settlements, subjugation of indigenous peoples and deportations of dangerous peoples by forcing them into becoming settler colonists. If governing metropoles meant subjugating dangerous populations by cruelty, it meant governing colonies by mass occupation, displacement and dispossession. As Foucault ([1976] 1978) expresses it, sovereign power is that which 'makes die and lets live'. We will see shortly how disciplinary power may have reversed this logic.

With the effective closing of national borders and mobility restrictions during the coronavirus pandemic we did not witness the 'return' of sovereign power. It has always been there but its exercise does not require the same strategies and technologies of power such as those noted above. Lockdown, curfew, confinement, regulation of movements, border controls and overall restrictions on the mobility of subject peoples are among the routinised and institutionalised technologies that sovereign power developed over a long period. What we witnessed was the deployment of new technologies of sovereign power during the coronavirus pandemic. Some polities may have been identified for acting most swiftly in closing borders and some others for shutting down not only external borders but also internal borders by enclosing entire cities. Yet let us not forget that both external and internal borders were subjected to immediate controls on all continents though with varying intensities across different states. These are examples of how the exercise of sovereign power had become routinised and latent and in turn less sensible and intelligible over time and yet ready to be reactivated to extract obedience from subject peoples. While accepted and even supported by dominant groups, and at the same time contested and brutally and cruelly experienced by others such as refugees, borders have become taken-for-granted forms of sovereign power. If the cruelties of borders as technologies of sovereign power have not been widely recognised that is perhaps because sovereign power has rendered itself less sensible and intelligible since its technologies have become routinised in thought and bodies. We thus ought not to conflate intelligibility with inexistence. Nor should we be surprised by the widespread obedience that sovereign power has extracted despite occasional and scattered protests primarily in the United States but also in other countries such as Germany and the UK questioning restrictions on movement mobility. What we saw during the pandemic was that sovereign power is entangled with other forms of power from which it draws strength but from which it needs to be analytically separated. Overall, a central insight from the collective experience of the coronavirus pandemic is that, unlike 500 years ago, sovereign power in the present could not have functioned without relying on disciplinary and regulative forms of power, which undoubtedly contributed to its concealment. What makes arguably sovereign power more effective is its reliance on disciplinary power and subjects who already embodied affects and effects of both disciplinary power and regulative power.

How to do things with disciplinary power

From about 200 years ago we witness the emergence of a new form of power that operates on the body: its discipline, its capacities, its will. The key object is the human body. It is true that sovereign power also operated on the human body: cruelties from branding black bodies to spectacles of torture and transportations (i.e. forced displacements) regardless of their deadly consequences were widespread technologies. For the accumulation of subject peoples required exercising sovereign power as the right to decide over the life and death of bodies.

Yet, as Foucault ([1976] 1978, 139) argues, starting about 200 years ago 'this power over life evolved in two basic forms; these forms were not antithetical, however; they constituted rather two poles of development linked together by a whole intermediary cluster of relations'. For Foucault disciplinary power was formed first and it was 'centred on the body as a machine: its disciplining, the optimization of its capabilities, the extortion of its forces, the parallel increase of its usefulness and its docility, its integration into systems of efficient and economic controls' (139). This was a form of power whose strategies and technologies involved optimising the capacities of bodies for production. It may have started in barracks (soldiers) and on ships (slaves) but disciplinary power gradually produced prominent assemblages where strategies and technologies of power and forms of knowledge combined to create optimised bodies for production in factories and workhouses. Over the following 200 years disciplinary power would produce clinics, prisons, workhouses, hospitals, schools, camps and eventually modern gyms, shops, studios and other assemblages where forms of knowledge were brought to bear on peoples governing themselves.

Just consider how we collectively became experts in the politics of our own bodies during the coronavirus pandemic. We not only followed daily what medicine had discovered about the virus and its modes of infection but also internalised injunctions and admonishments on how to conduct ourselves safely for others. We were advised to sacrifice everyday activities by isolating in order to save ourselves, others and public health systems. We developed, in an astonishingly short period of time, new forms of conduct by protecting ourselves and others in physical distancing, covering our faces and regulating our contacts. We developed astoundingly ritualised hygiene practices of disinfecting ourselves. We exercised all these forms of submission that disciplinary power calls for as subject peoples concerned with our own and each other's health and safety. If we followed the rules of confinement imposed by sovereign power obediently, we followed the rules of safety called for by disciplinary power submissively. What the pandemic rendered visible is that we experience these two forms of power simultaneously. We, our bodies, recognised how these two forms of power – sovereign and disciplinary – depend on each other and work together. Under normal circumstances neither form of power is sensible or intelligible. Under the pandemic situation they become revealed. Without a hint of irony, for those who needed help, practical guidance was offered on how to relearn socialising after the confinement.

If Foucault's analysis of Jeremy Bentham's panoptical prison design captures this relationship between sovereign power and disciplinary power, which was governed by discipline and punishment, the fines, charges, attestations, permissions and identity cards mobilised to separate those who were successfully responding to sovereign power by exercising discipline from those who were not during the pandemic illustrated it. Foucault explores how disciplinary power slowly comes into relation with another form of power that informs it. It is this third form of power that troubled Foucault in the late 1970s and the subsequent studies on power since. Foucault originally designated it as 'biopower' and its associated exercise as 'biopolitics'. To address some of this trouble we prefer to call it a regulative form of power for reasons we explain below.

How to do things with regulative power

For the last 250 years or so, a power that is at once totalising and individualising has emerged. It is regulative power 'focused on the species body [population], the body imbued with the mechanics of life and serving as the basis of the biological processes: propagation, births and mortality, the level of health, life expectancy and longevity, with all the conditions that can cause these to vary. Their supervision was affected through an entire series of interventions and *regulatory controls: a biopolitics of the population*' (Foucault [1976] 1978, 139, emphasis original). This form of power is regulative and not merely regulatory because it involves not only juridical-political regulation but also practical and habitual adjustment. Foucault observes that the emergence of 'this great bipolar technology – anatomic and biological, individualising and specifying, directed toward the performances of the body, with attention to the processes of life – characterized a power whose highest function was perhaps no longer to kill, but to invest life through and through' (139). This enabled Foucault to see a key relation between disciplinary and regulative forms of power. Each depends on the other but now disciplinary power functions most effectively as a positive rather than negative force. While Foucault never uses this term, we think it is quite appropriate to define this interdependent adjustment as calibration (Isin 2004). Foucault's discovery was to see a new relation between species and bodies. If disciplinary power was concerned with the optimisation of the body for specific purposes such as military or productive pursuits, it adjusted bodies to their demands. What regulative power performs is a different strategy of calibration: it derives the formulation and/or prescription of appropriate forms of conduct for bodies that are necessary for, or conducive to, the functioning of a population's health and wealth. More importantly, regulative power calibrates the conduct of bodies with that of a population *not* by admonishing or punishing bodies for non-compliance (though that relationship between sovereign power and disciplinary power continues to function) but by persuading, guiding, nudging and cajoling bodies that their health and wealth derives from it. Bodies discipline themselves as responsible subjects for their own and for the common good of the population of a polity. This is why Foucault increasingly saw sexuality as an apparatus of government as a primary example of calibration between the needs of a population (fertility, mortality, longevity) and the demands on bodies. To exercise regulative power over a population is impossible unless its desirable effective rates are converted into demands on bodies that compose it: fertility and mortality must be translated into forms of conduct conducive to achieving those effective rates.[18]

The metric that became a symbol of the coronavirus pandemic, the reproduction or R number, illustrates regulative power quite well. As explained by authorities, R is the rate by which each body infects other bodies or the virus reproduces itself. If a given body infects three bodies, the reproduction is three times higher than if a body infects only one other body. The logic of calibration here is that, if the body in question is identified, sequestered and isolated, its harm to the population is neutralised. One UK government advertisement showed the R-rate with a speedometer-like graphic indicating the-then current rate of infection and admonishing people to 'stay alert to keep R down'. Once epidemiology performs its function to calibrate bodies to populations, medicine can perform its function to cure the individual body and invest in its life. Much was made initially about the concept of herd immunity that would be gained by large numbers of people contracting and then recovering from the coronavirus. What is herd immunity if not essentially the exercise of the sovereign right to decide life and death of peoples, especially when it eventually became clear that the elderly, the infirm, the poor, indigent, black and brown bodies most disproportionally lost their lives? If sovereign power 'makes die and lets live', as we mentioned above, regulative power 'makes live and lets die' (Gros 2015).

To return to the relation between the accumulation of subject peoples and accumulation of capital (or between health and wealth), we have witnessed a tension during the coronavirus pandemic in terms expressed as the trade-off between health and the economy. When does sovereign power (re)start the economy? What is the trade-off between lives and livelihoods? If indeed the accumulation of subject peoples engenders accumulation of capital and accumulation of knowledge, an analysis of forms of power must keep all these three processes in view as they are intertwined. Moreover, just as there are different forms of knowledge and subject peoples, capital must also be understood in its different forms (economic, cultural, symbolic), as Pierre Bourdieu ([1979] 1984) insists.

We may have overlooked how the overlaps and dynamics between different forms of power function and how each depends on certain aspects of the other in the coronavirus pandemic. Our aim here is to both provide an historically informed overview of forms of power and their simultaneous existence and how the development and articulation of a new form of power has increased their complexity.

We recognise that questions of power and especially biopower have given rise to disagreements on its functions, effects and transformations. In particular, the concepts of biopolitics and biopower have been expanded by Ian Hacking (1983), Giorgio Agamben (1998), Michael Hardt and Antonio Negri (2000), Roberto Esposito (2008), Nikolas Rose (2006), Thomas Lemke (2011) and Achille Mbembe (2019). As Paul Rabinow and Nikolas Rose (2016) have shown, however, each scholar has taken 'biopolitics' in a particular direction and with mixed results. Rabinow and Rose insist that if biopower and biopolitics must retain their analytical power we must include at least three elements: forms of knowledge about life; strategies that intervene in the name of life; and modes of subjectivation through which people invest in their own lives. Taking simultaneously totalisation and individualisation as their key analytical tool they illustrate how biopower functions by regulating between bodies and populations. This is broadly how we see regulative power but the terms biopower and biopolitics, beginning with Foucault's explorations, have conflated the relations between sovereign power and disciplinary power and between disciplinary power and regulative power and generally overlooked to separate power (in) theory from power (in) practice. As Paul Patton's (2018) analysis shows, it is very difficult to imagine how biopolitics intervenes at the level of populations rather than through individual bodies without some mechanisms between the two. Frédéric Gros (2015) astutely shows that Foucault shifted his attention to studying government precisely because of the difficulties of separating disciplinary power from regulative power and of identifying mechanisms of calibration between bodies and populations. Moreover, there is also the problem of trying to fit our current experience into biopower and biopolitics when a new form of power has emerged that requires articulating new concepts.

To put it differently, interpreting and evaluating the developments in the last forty years through the analytics of biopower when a fourth form of power was emerging creates problems. While studies on modern power have been attempting to plot various new events such as the development of apps, devices and platforms into a historical series already named and recognised (Amoore 2015; Rouvroy and Berns 2013), a new event may have been unfolding in the present but, like we stated earlier, a new form of power perhaps remained dimly sensible and barely intelligible – until the coronavirus pandemic. This is why we attempt to name a new form of power historically that in turn will enable us to interpret post facto the three forms of power. We shall explore the birth of datasciences distinct from but related to statistics and computing in the last forty years in the next chapter and the birth of neurosciences in the following chapter but now we return to the coronavirus pandemic to read the signs of the birth of sensory power. What were these signs?

The birth of sensory power

The key in the invention of sensory power has been the objects that are constructed between bodies and peoples. The birth of sensory power is not as bipolar as regulative power as Foucault identified: individualising and specifying, anatomic and biological or molar and molecular. Although governing bodies with attention to the processes of life necessitates segmenting populations into what Hacking (2007) calls 'kinds' of peoples, sensory power adjusts or calibrates peoples into polities in ways that are much more dynamic and fluid than regulative power. Foucault ([1976] 1978, 139) anticipated that disciplinary and regulative forms of power were 'linked together by a whole intermediary cluster of relations' but what were these clusters? What are these kinds of relations between bodies and peoples? How does sensory power assemble the kinds of people? For regulative power and disciplinary power, populations were produced through categories such as class, gender and race as partitions of a population. For sensory power clusters may partition populations but its logic does not depend on preconceived categories such as fertility, mortality and longevity. The idea of clusters was barely sensible and intelligible forty years ago when Foucault and Deleuze wrote about power. Since then and especially during the pandemic we have witnessed a new form of power that assembles clusters as transversal objects of government that are based neither on categories or classifications but are dynamic and fluid objects of interests for specific goals and purposes of government. These clusters are brought into being through sensory assemblages which integrate apps, devices, peoples and platforms. These are sensory assemblages in the sense that they are based on data gathered from sensory apparatuses built into these apps, devices, platforms and indeed peoples and things.

The word 'clusters' may be thought to have relevance only for the coronavirus pandemic and the epidemiological models and vocabularies that dominate public health discourse. As we shall see, as a concept, practice and method it is a key object for sensory power. Let us recall that Foucault originally developed his concepts of disciplinary and regulative forms of power with meticulous attention to responses to eighteenth- and nineteenth-century epidemics and illustrated that languages of power were deeply entwined with them (Foucault [1975] 1977, 195–9). It is necessary to understand this object in activities of government before we trace its emergence in forms of knowledge as a method signalling a transformation from statistics to analytics also leads to machine learning models.

How do we observe sensory assemblages that bring clusters into being? The sensory assemblages will appear resonant with Deleuze's ([1986] 1988, 32–41; [1980] 1987, 1837, 1227) interpretations of Foucault and subsequent developments. Paul Patton (2018) explores Deleuze's ([1990] 1995b; [1990] 1995a) suggestions of societies of control. Leaving aside assumptions that we have already questioned – that societies of discipline were giving way to or were replaced by societies of control or that biopolitics intervenes without intermediary mechanisms – Patton nonetheless makes pertinent observations on how control societies were oriented to technologies of modulating bodies rather than punishing or disciplining them, and in doing so were creating new assemblages. The recent scholarship on data politics inspired by primarily Foucault and Deleuze show that these were prescient observations (Amoore 2015; Beckman 2020; Galloway 2006; Mackenzie 2017). However, interpreting 'control societies' more like a continuation of 'disciplinary societies' governed through new technological means such as biometric recognition, automated surveillance, algorithmic government and digital monitoring largely overlooks the birth of a new form of power. That sensory power is a related but a different form of power from what control has come to mean in the phrase 'control societies' is important to recognise.[19]

The objective is, however, not to propose the dawn or age of 'sensory societies'. The difficult task ahead is to consider these studies on datapolitics by resignifying them through analytics of power reworked here, and then to explore how a fourth form of power is nestled in but became sensible and intelligible among the other three forms of power during the coronavirus pandemic. Let us now specifically consider how the coronavirus pandemic has rendered sensory power sensible and intelligible.

All forms of power work through assemblages that enact their objects through strategies and technologies: sovereign power governs territories through assemblages that enact colonies, dominions, states (cartography, maps, surveyors, borders); disciplinary power governs bodies through assemblages that make up prisons, camps, hospitals, factories, prisons, schools, workhouses (architecture, walls, fences, guards, gates); and regulative power governs populations through assemblages that enact attributes, categories and classifications such as class, gender and race (administrative records, enumerations). Sensory power governs clusters through new assemblages of apps, platforms and devices (code, sensors, transmitters and protocols). These assemblages are sensory in the sense that they operate through and act upon sensory data.

While sensory power works through sensory assemblages to produce its objects, it is also articulating with sovereign, disciplinary and regulative forms of power. Thus, a new form of power is to be found nestled in existing forms but nonetheless deploying new strategies and technologies that are becoming sensible and intelligible. We can now highlight clusters as objects of government that sensory power assembles and some of the concepts, practices and methods that it produces.

Hotspots, bubbles, clusters

We have seen reference to various entities such as hotspots, bubbles and clusters. These were of course related to the spread of the coronavirus but they were meant to indicate how bodies either infected or healthy were related to each other. What made them hotspots, bubbles or clusters is that these bodies came into contact with other bodies with a condition (infectious, healthy) that made them an object of interest for government. This interest is not about capturing, punishing or disciplining these bodies but about identifying and providing the means by which bodies can modulate their behaviour or conduct with desirable outcomes. Modulation is different from the technologies of adjustment of regulative power, which works through forms of knowledge arising from frequencies, trends, periodicities. By contrast, governing clusters involves ongoing and 'live' tracking of their performance through which decisions and interventions can be formulated with immediacy by narrowing the time between identification and intervention. This difference can be exemplified by considering how gatherings of bodies for celebrations, raves and parties elicited new policing concerns about discipline and punishment during the easing of the lockdown between 2020 and 2022. By contrast, clusters as objects of sensory power do not simply exist as spatial and temporal gatherings of bodies but are relations between 'infectious' and 'healthy' bodies that nevertheless develop spatial and temporal forms. The sensory assemblages that produce clusters involve relations between human and non-human actors including devices, sensors, platforms, practices, data, and methods and agencies, authorities, technicians and professionals in governments, corporations and non-governmental organisations. The devices that make up sensory assemblages are not entirely digital as contact tracing programmes involving public health personnel and analogue practices illustrate. Tracking, testing and tracing bodies require continuous gathering, storage and transmission of analogue and digital data by various agents and authorities. Along with lockdown, distancing and isolation, tracking subjects who have been infected, tracing subjects who may have had physical contact with them, and alerting both clusters to isolate themselves requires exercising technologies of power appropriate to these objectives.

The tracing and tracking apps we have seen involve competitive struggles between and among projects and professions involving states, international organisations and multinational corporations to develop knowledge about objects and subjects of interest. The pandemic has shown how technology companies command such knowledge and intensely compete with each other for procuring projects. But the competition is also between various competing professions involving epidemiologists, statisticians, data scientists, programmers, app developers, security experts, methodologists, financiers, investors and so on who are transnational and whose expertise traverse national borders. While sensory assemblages may not be entirely digital, they nonetheless involve infrastructures of digital technologies such as satellites, data centres, transmitters, receivers and mobile devices and include analytics such as algorithms, machine learning and cloud computing. Through global connections between human and non-human actors such as devices, engineers and programmers, these projects and professions accumulate data about infections and deaths, which in turn inform their development of tracking and tracing apps. Here we can see how the tracking and tracing performed by these major technology corporations are by no means limited to disease but related to other forms of conduct such as watching, listening, reading, communicating and so on and through which people form relations that can be produced as clusters. While producing and controlling clusters to maintain the accumulation of subject peoples (health) and accumulation of capital (wealth) has proved an elusive objective, the development of coronavirus apps has made visible a form of power whose object is the production, identification and management clusters. That is, clusters are not novel to coronavirus as objects and subjects of power but related to sensory assemblages already in operation in several fields of commerce and government.

Transversing clusters

If sensory assemblages propagate, multiply and reproduce clusters as objects and subjects of interest for governing authorities through projects and professions how are they constructed? Clusters do not merely constitute 'new' representations of 'old' populations. Clusters do not map on to populations of regulative power mostly compiled for and by national and corporate authorities. Within its first six months the coronavirus traversed borders and reproduced itself across nearly 200 states despite closing national borders with variable intensities. What is lost in cross-national comparisons is how these states were only marginally able to control the transversal reproduction of the coronavirus. At the same time, sensory assemblages spread, multiplied, adapted and mutated clusters. To understand this requires seeing how sensory assemblages produce the objects and subjects they represent. It is for this reason data can be understood as an agent within sensory assemblages for what they might perform depending on dispersed, distributed and transversal relations. When data about clusters are brought into being by sensory assemblages of experts, methods, technologies, organisations, practices and authorities, it is never under the strict control or influence of any of them. It is through its circulation that sensory assemblages get detached from the authorities that make them up and govern objects as well as subjects in complex ways. It is not only clusters of the infected and those who are contacted by the infected that traverse borders but also different combinations of strategies and technologies, knowledges and all of the actants and actors that constitute multiple clusters (networks, associations, relations) that invariably traverse the borders of cities, states and empires.

Sensible clusters

If indeed conventional population statistics typically involve sociodemographic categories, it often involves the collection of data through self-elicited accounts that use various methods to fit people into categories.[20] As Hacking (2007) observes, the making of these categories produced rigid new conceptualisations of the human being. By contrast, sensory assemblages construct clusters based on the performance of bodies rather than imposing categories and classifications in advance. Unlike categories, clusters are generated as a consequence of analytics such as machine learning and algorithms that do not identify associations between existing variables but explore multidimensional patterns among 'hundreds and in some cases tens of thousands of variables and sample sizes of millions or billions of data' (Mackenzie 2015, 434). The differences are not understood as statistical variables but derive from combinations of attributes from different 'forms of data (text, images, video, transactions, sensors), not just the variables measured using classical statistical tabulations of surveys, polls or random sampling' (Mackenzie 2015, 433). While variables can also be diverse, a key difference from the conventional production of populations is the registering of multiple forms of conduct or what people do such as their movements and actions (transactions, choices, statements, interactions) where the objects and subjects of interest are not the individual factors that affect conduct, but on aggregate patterns and connections: contagion, dissemination, influence, association etc. It is through the continuous tracing, tracking, monitoring and modulating work of sensory assemblages that produce clusters that are fluid (and dynamic) rather than solid (and static) objects of government. If clusters are rendered sensible as fluid and dynamic, how are they rendered visible?

Intelligible clusters

Making clusters intelligible is not a question of representing them but it is part of their production. Each form of power produces its regime of visualisation, from cartography to anatomical diagrams and statistical charts, but the visualisation of sensory assemblages has precipitated entirely different technologies of production. If Edward Tufte ([1983] 2001) famously insists that visualisation can be a technique 'for reasoning about statistical information' (9) that 'reveals data' and can be more 'precise than conventional statistical computations' (13), it is Stephen Few (2006), also famously, who operationalises its logic: dashboards. The millions of data points that make up sensory assemblages have mobilised visualisations as not simply representations but a key method through which data scientists make data sensible and intelligible (Mackenzie 2017, 437). These visualisations identify unseen patterns and include interactive elements and dashboards that enable seeing the effects of combining different data on features of a population. That the 'dashboard' has become a ubiquitous technology of government like cartography, anatomy and charts is a sign of sensory power. We will discuss in Chapter 4 how it is used in governing cities. For all the seemingly accurate cartographic representations that dominate publicly available visualisations about the coronavirus pandemic such as those showcased by various authorities, often all that they have offered, especially in the early years of the pandemic, are fairly basic data and statistics mapped onto national borders. There are more sophisticated 'dashboards' such as those for financial systems (markets, transactions), transportation systems (air, rail, sea), military operations (theatres) and managing football games (Mattern 2015). There are also dashboards in fields such as migration or policing, where dashboards have been deployed for governing movement and criminality (Tazzioli 2020). While initially most city dashboards used previous data generated in specific periods, they increasingly came to incorporate data produced in real time by sensors and devices including data scraped from social media and through crowdsourcing.[21] So, while dashboards on the pandemic were relatively basic, they are part of intensive investments by governments, research funders and universities that are underway, and they are developing into more sophisticated forms of visualisation as part of production of clusters as the production of objects and subject of interest. Such dashboards increasingly resemble control rooms designed for performing sensory power. This brings us to the most important aspect of clusters: they are considered live objects and subjects of interest for government.

Pulsing clusters

While data may be presented in dashboards by disciplinary or regulative forms of power, sensory power organises machine learning algorithms so that measurement, identification, action and intervention can happen simultaneously, instantaneously and continuously. The objects and subjects of interest for governing become live with temporal and spatial intensities.

The fluidity of clusters makes them live but also renders them difficult to control by any singular authority. It is this aspect of sensory assemblages that brings us closest to what we mean by sensory power. It involves modulating the performance of bodies and clusters through governing interventions that rely on technologies of machine learning, algorithms and visualisations of clusters as relations. Rather than the periodic 'stocktaking' of conventional statistics, populations are partitioned into clusters that are live and have pulses, flows and patterns. Rather than the accuracy and precision of data on numbers infected, contacted, recovered and deceased, it is knowledge of the spread, peaks and troughs of the disease that mobilises governing interventions. In turn, data serve a dual function: for identifying attributes or features that make up clusters (e.g. infected, contacted) but then monitoring and evaluating those features live (e.g. daily changes in R metric hotspots, bubbles and clusters), thereby provoking changes in conduct (how people govern their movements) and sovereign and disciplinary interventions (easing or increasing lockdown). Identifying attributes produces data in much the same way as classical data regimes: populations are periodically measured with indices, rates, metrics and indicators. However, modulation between bodies and populations through clusters enacted by sensory assemblages works continuously with pulses and signals. It is a recursive logic captured in four stages of data analytics: descriptive and exploratory analysis (what is happening, often in real time); doing predictive analytics ('what is likely to happen'); detection ('tracking who is likely to succeed and who will fail'); and evaluation and data diagnostics ('how to improve programme performance').[22] What these elements encompass is how the data of sensory power is not separate from but interwoven with sovereign, disciplinary and regulative strategies. The coronavirus pandemic made especially this aspect of sensory power sensible and intelligible.

The attempts to develop an app to track and trace people, warn if they are near infected people, and ask to change their location were perhaps most revealing. The app aims at *tracking* the locations of bodies infected with the virus, *notifying*, *testing* and (if necessary) *isolating* them in order to stop its reproduction, *tracing* all bodies that infected bodies came into contact with, *notifying*, *testing* and (if necessary) *isolating* them as well, and thus *slowing* the reproduction (R-value) of the virus. Essentially this creates a live cluster of bodies infected or potentially infected by the coronavirus. Governing bodies in clusters, however, requires interventions at the stages of notifying, testing and isolating in order to be effective. Clearly, this is a relation between regulative power and disciplinary power: to achieve the desired infection rate R, disciplinary technologies of power such as the consent of individuals to agree to be notified and act according to the results whether that involves getting tested or (if necessary) self-isolating. This is a costly and inefficient exercise of power. Yet the enthusiasm about a potential app and its promise to deliver a game-changer is palpable: to minimise disciplinary power and instead maximise sensory power, in other words to formulate the problem of government as a relation between regulative power and sensory power.

Such an app never properly functioned as intended despite numerous attempts. But its various stages of the cycle – tracking, notifying, testing, isolating, tracing, notifying, testing and results – illustrate how clusters are formed as pulsing objects. If solutions had been found to automate the testing and isolating stages, essentially multiple, relational, fluid, visualised and live clusters could be *governing themselves*. There are of course technological limits to such a scenario. There are also severe legal, political and cultural limits but such limits may become surmountable soon in other fields of application where such limits seem less relevant such as in finance or transport logistics.

We now know that numerous practices for tracking and tracing peoples are already present in fields other than epidemiology. We mentioned earlier that the accumulation of capital in finance, manufacturing, retail, transportation, hospitality, entertainment and other industries increasingly rely on the accumulation of subject peoples through tracking and tracing their movements and the modulation of sentiments, needs and desires. We have also mentioned finance, policing, crime, migration, borders and education as such fields of government where sensory power is making its appearance. The live data produced from sensory assemblages pervade these sectors and fields. The coronavirus pandemic may have accelerated them but strategies and technologies of sensory power have been emerging over the last forty years in these fields.

The range and volume of data about bodies and things being generated through the internet and devices and sensors by organisations, agencies, corporations and governments produced new concepts (correspondence), practices (mining), methods (clustering and learning), subjects (datascientists, neuroscientists) and objects (clusters). The implications of these developments have been studied in relation to privacy, anonymity, security, legality and other concerns that are cast in terms appropriate to modern polities everywhere. As we have argued above, just as metropole–colony relations were configured through the invention of regulative power, we expect that new developments in datasciences are reconfiguring not only metropole–postcolony relations but also sensory power. In other words, not only do issues such as privacy and security play out differently in the metropole than in the postcolony, but the agents of power-knowledge and the targets of government are also constituted differently. Moreover, datapolitics plays out differently in the metropole and the postcolony precisely because of the different ways in which certain rights are protected, developed or violated.

If we turn our attention away the pandemic and towards the development of datasciences (Chapter 2) and neurosciences (Chapter 3) and return to the city (Chapter 4), we will see how a long history of power provides us the workings of sensory power.

Notes

1. We are drawing from Spinoza (2016; 1985), Nietzsche ([1887] 2007), Deleuze ([1969] 1988; [1986] 1988; [1962] 1985) and Foucault ([1971] 1977; [1976] 1978; 1982; [1984] 1997) on power through Austin (1962; [1961] 1970) and Derrida ([1977] 1988).
2. There are numerous occasions where Foucault makes this statement but perhaps most emphatic is: 'To put it bluntly, I would say that to begin the analysis with a "how" is to introduce the suspicion that power as such does not exist' (Foucault [1982] 2000, 336). And Foucault was clear that '"How?" [was] not in the sense of "How does it manifest itself?" but "How is it exercised?" and "What happens when individuals exert (as we say) power over others?"' (337).
3. We find Foucault making these connections throughout his work but especially in his 1970–1971 and 1981–1982 lectures and interviews (Foucault [2011] 2013; [1982] 2000). We also find him almost always in dialogue with Spinoza, Nietzsche and Deleuze in various periods. We also find Foucault in dialogue with Austin, especially in his early writings, though Austin remains unnamed (Foucault [1969] 1972, 93; 2019a, 47).
4. As Foucault was writing through a movement of thought that indeed revolutionised speaking about power by analysing its affects and

effects, we also write through that movement which includes Spinoza, Nietzsche, Deleuze, Derrida, Bourdieu, Rancière and Austin. Deleuze ([1962] 1985; [1986] 1988), and Foucault ([1975] 1977; [1976] 1978) are explicitly or implicitly but always in dialogue with them. It may be surprising to find Austin in this list, as Deleuze and Foucault often distinguish their thought from Austin's. But this is also how opponents become influential in the invention of concepts. We certainly see Austin as part of this movement of thought.

5. As Deleuze would say about Nietzsche's relationship to Spinoza and his own relationship to Nietzsche, these are not master–disciple relationships based on debt but generative and transmissive relationships of critical thought. 'But more than one writer before Nietzsche had spoken of a will to power or something analogous; more than one after Nietzsche spoke of it again. . . . But the latter were no more disciples of Nietzsche than the former were his masters.' (Deleuze [1962] 1985, 80).
6. 'Every relationship of forces constitutes a body – whether it is chemical, biological, social or political. Any two forces, being unequal, constitute a body as soon as they enter into a relationship' (Deleuze [1962] 1985, 40). 'The will to power is thus added to force, but as the differential and genetic element, as the internal element of its production. It is in no way anthropomorphic' (Deleuze [1962] 1985, 51).
7. 'We can see that the problem of measuring forces will be delicate because it brings the art of qualitative interpretations into play' (Deleuze [1962] 1985, 42).
8. 'Every force is related to others and it either obeys or commands. What defines a body is this relation between dominant and dominated forces' (Deleuze [1962] 1985, 40).
9. Deleuze ([1962] 1985, 50), emphasis original.
10. 'Forces can only be judged if one takes into account in the first place their active or reactive quality, in the second place the affinity of this quality for the corresponding pole of the will to power (affirmative or negative) and in the third place the nuance of quality that the force presents at a particular moment of its development, in relation to its affinity' (Deleuze [1962] 1985, 61).
11. 'But more than one writer before Nietzsche had spoken of a will to power or something analogous; more than one after Nietzsche spoke of it again. . . . They always spoke of it in the sense expressly condemned by Nietzsche: as if power were the ultimate aim of the will and also its essential motive. As if power were what the will wanted' (Deleuze [1962] 1985, 80).

12. Foucault would make this point as early as 1970: 'If, in Western philosophy until Nietzsche, desire and will were never able to rid themselves of their subordination to knowledge, if the desire to know was always doubled by the precondition of knowledge, the reason is this fundamental relationship to truth. That is why we can say that Spinoza had already gone to the furthest limit: at the highest point of this theoretical structure, at the point where he was closest to getting outside it and overturning it' (Foucault [2011] 2013, 24). And 'Spinoza names in plain language, and in order, what makes classical metaphysics possible. Now it is in fact this will–knowledge–truth relationship that is at stake in Nietzsche. As a first approximation, Nietzsche's texts can (and should) be read as an attempt to free the desire to know from the form and law of knowledge' (Foucault [2011] 2013, 25).
13. Thomas Lemke (2021) suggests that Foucault's concept of government was meant not only for people but also for things.
14. The history of modern computing has even longer history, going back to the 1690s. There are many impressive histories of computing but almost all provides what might be called 'internalist' histories. See, for example, Haigh and Ceruzzi ([1998] 2021). An exception is Jon Agar (2003). For histories of the brain that avoid 'internalist' temptations, see Malabou ([2004] 2008), Pitts-Taylor (2016), Rose and Abi-Rached (2013) and Smail (2008).
15. Thomas Lemke (2011) provides an introduction to how Foucault's conception power as life-giving allowed him to distinguish sovereign power from regulative power and how some of his interpreters blurred them.
16. We have developed apparatuses of government and how they are assembled with an illustration of citizenship as an apparatus of government elsewhere (Isin 2024).
17. Stuart Elden (2016; 2017) and Thomas Lemke (2019) provide most helpful accounts of how Foucault's lectures and books coincided or diverged. They offer insights into especially the lectures *Society Must Be Defended* ([1997] 2003), *Security, Territory, Population* ([2004] 2007) and *The Birth of Biopolitics* ([2004] 2008). We cannot enter into this complicated matter in the space of this book but we are aware that these lectures address some of the four disagreements we noted above.
18. Foucault makes his interest in sexuality clear by stating that 'sexuality exists at the point [251] where body and population meet. And so it is a matter for discipline, but also a matter for regularization' (Foucault [1997] 2003, 252). This point where body and population meet can

also be different apparatuses of government such as security, citizenship, health or productivity.

19. We are drawing from Bigo, Isin, and Ruppert (2019, 1–17).
20. Libby Schweber (2006) provides how the production of vital statistics in France and England in the nineteenth century involved such an assemblage. Similarly, Evelyn Ruppert and Stephan Scheel (2021) show how European population as an object and subject was brought into being in the twentieth century.
21. Kitchin and McArdle (2018); Kitchin, Lauriault and McArdle (2015b) and Mattern (2015).
22. These stages are explored in more detail elsewhere (Isin and Ruppert 2019).

2

Doing things with numbers: datasciences

To distinguish between power as such and power as performatives and to understand the invention and development of different forms of power as performatives (sovereign, disciplinary, regulative and sensory) requires exploring the corresponding forms of knowledge, comprising a heterogeneous group of concepts, practices and methods and how each group is intertwined with power as they produce objects and subjects of knowledge. To say that forms of power are intertwined with forms of knowledge does not mean that there is a symmetrical or synchronous relationship. Nor does it mean that they are identical, as in saying 'knowledge is power'. As Alain Desrosières suggests, the history of statistical reasoning emerges from the relationship between technical and social objects, and statistical objects are interwoven with them.[1] Ian Hacking similarly observes that the erosion of determinism and the taming of chance in the history of probability were preceded by the inventions of disciplinary and regulative forms of power. Hacking insists that the erosion of determinism was related to the changing ways of governing people and did not necessarily create freedom but rather more possibilities for constraint, which created the challenge of the taming of chance for both power and knowledge.[2] The different forms of knowledge and their affects and effects produce complex forms of power and we need to understand their relations.

If the body and its capacities, drives and conduct have become a different type of object of the datasciences in the twenty-first century, and if sensory power is the driving force for this development, we must revisit how the datasciences came into being. Where have their strategies and technologies been developed and experimented with? We want to explore how the body, as an object of datasciences, focuses on its movements and conduct precisely with the data collected by mobile and immobile devices equipped with sensors and how the image of the body undergoes a transformation. The assemblage of datasciences has come to constitute the body in a different way since the 1930s. How the development of this form of knowledge is intricately associated with the birth of sensory power and its strategies and technologies is the key to understanding how datasciences and neurosciences have converged since the 1930s and accelerated in the 1980s. That the datasciences have become a site of social and political struggle in the twenty-first century not only because it concerns anyone who is connected to the internet but also because it reconfigures relationships between states, subjects and citizens by creating novel polities is an important aspect of this convergence. It is now obvious to almost everyone that every device is now connected to the internet and generates vast quantities of digital data about interactions, transactions and movements whether users are aware or not. From its origins as a military network the internet rapidly became the space over and through which governments and corporations began collecting, storing, retrieving, analysing and producing data that analyse what people do and say on the internet. This ranges from who communicates with whom, who goes where and who says what – and much more besides. This is now being augmented with data that people produce about themselves, especially their relations, their movements and sentiments; the amount and range of data that have become available is, as everyone now knows, staggering. An historical ontology of the object we shall call 'polities as learning machines' requires outlining a history of datasciences associated first with disciplinary power and then with regulative power in the eighteenth and nineteenth centuries, and how the development of neurosciences converges with datasciences in the twentieth and twenty-first centuries by transforming brain–machine analogy into machine intelligence.

The ways in which knowledge is produced through power relations require an examination of the role of language, official and unofficial representations and the distinction between discursive and non-discursive practices. The ways in which power and knowledge are mutually constitutive generates complex strategies and technologies. The historical specificity of knowledge, with different modes of objectification that interpellate people into becoming certain kinds of people, and the role of power in the possession and use of knowledge are related to struggles over different forms of power. If Michel Foucault's work on knowledge challenges dominant concepts of power, it is because he criticises the ideology of a separation between knowledge and power and seeks to show how power is intertwined with knowledge. Examining even one form of knowledge practice, such as that of the police, shows how a form of knowledge is deeply embedded in relations of power. If Foucault is interested in the concept of subjugated knowledges, which are forms of knowledge that have been disqualified or marginalised, it is because he is convinced that they reveal themselves to an analysis of the power relations that produce objects and subjects of knowledge. Foucault rightly insists that there is a relationship between will to power and knowledge, and that knowledge is an invention that is not inherent in human nature or the structure of the world. He highlights the complex and often complicated ways in which knowledge is produced and used in exercising power. Every field where power is exercised is a field where knowledge is formed, and every field of knowledge enables the exercise of power. The practices such as measurement, inquiry and investigation have all been both means of exercising power and methods for producing knowledge. The activity of government is inseparable from the activity of thought, and the analysis of the history of government requires attention to the conditions under which it becomes possible to hold certain things to be true about human beings and their interrelations.

For exploring the emergence of what we call 'datasciences', we need to illustrate how this assemblage of heterogeneous elements comes together, how the history of each of these elements and their relationship to each other shape datasciences as a form of knowledge, and how sensory power becomes both instrument and effect of it. Concepts such as surveillance, classification, correlation and correspondence; practices such as survey, census, inquiry and mining; and methods such as statistics, analytics, clustering and learning are all examples of such heterogenous elements that were invented and transformed through different forms of power from sovereign, disciplinary and regulative power and created the conditions of possibility of sensory power. By reworking these concepts, practices and methods we will resignify them now that we are making a new form of power – sensory power – sensible and intelligible. To put it differently, concepts, practices and methods that we have become familiar with by studying them in relation to sovereign, disciplinary and regulative power must be now reinterpreted and revaluated with our knowledge of sensory power.

While 'data' has been central to heterogenous elements coming together to constitute datasciences, we need to understand data as gathered, stored, tabulated and analysed – in other words, as a constituted object of knowledge. This process of making data is as heterogenous as forms of power and forms of knowledge that generate it. To put it differently, when data are claimed as the systematic collection and presentation of information in order to infer the relevant causal structure of complex phenomena, it simultaneously stakes out a domain of knowledge and a domain of power. It also institutes the

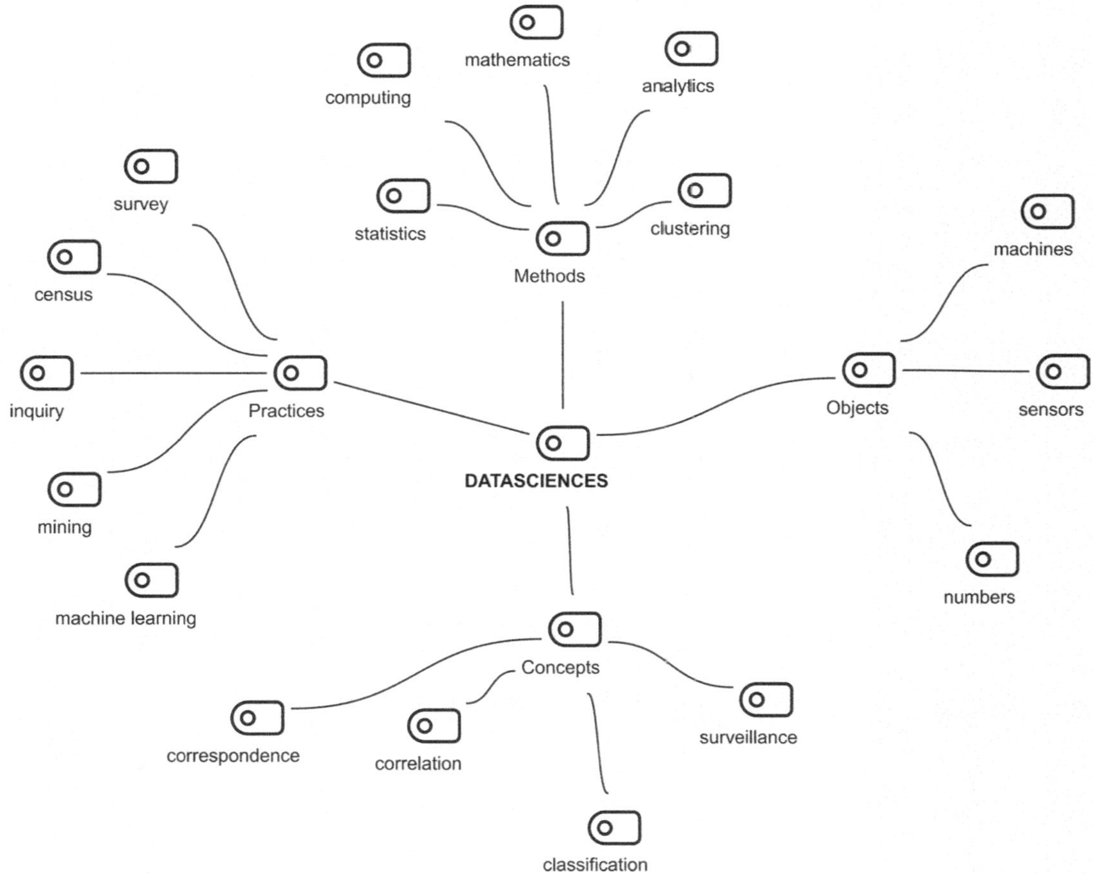

Figure 2.1 Datasciences or doing things with numbers

legitimacy of datasciences as a form of knowledge for extracting information from data in various fields and its potential for advancing research and solving problems.[3] Datasciences, as an assemblage of heterogeneous elements involving mathematical statistics, computer science, and their production and identification of persistent relationships between variables, produce significant affects and effects of power as performatives. The invention of the body as both object and subject of power historically had heterogenous trajectories with the invention of data.

The invention of population game in this chapter and the imitation game in the next chapter are two transformative moments in the convergence between datasciences and neurosciences. If the population game has resulted in the invention of an abstract object calibrating bodies and populations with its regulative principles, the imitation game has resulted in developing intelligent machines as polities with self-regulative properties.

The population game

Ian Hacking (1983) describes the period between 1820 and 1840 as the 'avalanche of printed numbers' in Europe and America. Hacking is reflecting on Michel Foucault's concept of biopolitics that targeted 'population' with its own characteristics as an object of government in the nineteenth century. This invention was related to developments, not least the birth of a science – statistics, Hacking's primary concern – but also associated sciences such as demography and epidemiology, and data practices such as surveys and censuses. Hacking emphatically characterises that as the period when the 'statistical study of populations comes to amass gigantic quantities of data' (280). Yet he recognises that the interest in population was mobilised by the interest in the body. If disciplinary power had discovered how to harness the body by both controlling and utilising it, regulative power provided a novel direction: the body was not only an object of government but also a subject that would govern itself to produce a healthy and wealthy population with security.

Although the contemporary period has been described as the era of data revolution, scholars place this in a much longer history (Mayer-Schönberger and Cukier 2013; Kitchin 2014). The personal computer of the 1980s developed into a pervasive device in the twenty-first century, became connected with other devices through the internet (a word born in 1980), converged storing and processing of data with sharing, led to the invention of protocols for collecting, representing and sharing of data and generated not only an internet of people but also of things. Now, the amount of data generated and collected from these devices and the interests, authorities and expertise required to render them useful make the data revolution of the 1820s appear rather minuscule but we need to understand that the invention of the population game spurred the data revolution between 1830 and 1848 and constitutes the origins of the invention of learning machines in the data revolution in the twenty-first century.

As Hacking was writing about 'gigantic quantities of data' in the 1980s, a new term was rapidly becoming popular in Euro-American languages: 'personal computer'. The invention of large-scale data processing machines following the Second World War was giving way to the miniaturisation of both processors and components of a computer – storage, graphics, controllers and cooling. By the late 1980s a personal computer could already store and process all the 'gigantic data' collected about populations between 1820 and 1840. This would have been wondrous to William Farr (1807–83), a compiler of abstracts for the newly founded Office of the Registrar-General of England and Wales (1836), who remained in office for forty years (Hacking 1983, 284). Hacking (1983, 291) recounts that it was Farr who pirated a Swedish computing machine with more than 5,000 moving parts for use in the Office of the Registrar-General. Two centuries later, a hand-held device could store and process such gigantic data.[4]

The discourse on the datasciences, emphasising their importance for understanding complex phenomena and developing intelligent algorithms for managing the multiplicity of people and things, has accelerated with the discourse on machine intelligence. This discourse includes not only concepts, practices and methods but also the objects and subjects it creates. As we have mentioned, we will use 'datasciences' to refer not only to the fields of knowledge involved, such as statistics and mathematics, but also to data practices, ranging from surveys, censuses, queries and mining and methods, such as statistics, analytics, clustering, algorithms and learning, in order to stage these together with the sensory power. We shall later see not only how machine learning techniques and machine intelligence are transforming the analysis of neuroimaging datasets but also how the construction of learning algorithms is producing new images of intelligence in humans and machines.

To provide an outline of the transformations in the concepts, practices and methods of what we now broadly call the datasciences, a generation of outstanding scholarship on the history of statistics, analytics and algorithms who studied how the production of these forms of knowledge relates to forms of power is necessary. Foucault and Bourdieu inspired this generation of scholars, who have taken up the relations between power, knowledge, subjects and objects and investigated the ways in which states, agencies, organisations, corporations and institutions – often assembled in different combinations as governing authorities – constituted their legitimacy and legality by producing knowledge about objects and subjects and by establishing concepts, practices and methods such as censuses, indexes, indicators, registers, rolls, catalogues, logs and archives. We now understand much better the relationships between state formation and statistics, probability and data regimes.

Why do we insist on bringing these multifarious and heterogenous elements under 'doing things with numbers'?[5] Why must critical thought investigate doing things with numbers?[6] The outstanding aspect of this literature inspired by Bourdieu and Foucault is that, despite noting numerous transformations in what we call here datasciences, it consistently aims to fit its findings into disciplinary and regulative power. While they present plenty of evidence making a case for identifying the birth of a new form of power, they fall back on disciplinary and regulative power. The development of concepts such as surveillance, classification, correlation and correspondence; practices such as surveys, censuses, inquiries and mining; and methods such as statistics, analytics, clustering, algorithms and learning provide this evidence.[7] Their historical and political trajectories, their role in shaping power relations, and their power to invent objects and subjects through concepts, practices and methods will show what datasciences as a form of knowledge assembled from the nineteenth to the twenty-first century activated by sensory power.

We may now recall what we called the genealogical principle in the preface and the introduction: while the objects we study may continue to have the same names, their function and operation are constantly changing. In discussing both the population game and the imitation game in the following two chapters, the question that guides us is this: given what we now know about sensory power, how do we see the development of its elements in the concepts, practices and methods that emerged from these games? We are not interested in repeating what we know about these concepts, practices and methods but in resignifying them from the perspective of sensory power in order to identify the conditions of their possibility.

Concepts: surveillance, classification, correlation, correspondence

The datasciences have emerged due to advances in sensing and storage technology, resulting in high-volume and high-dimensional data sets.[8] Generally speaking, datasciences involve extracting knowledge from data through computational science, statistics, data analysis, machine learning and pattern recognition. The scientific process in datasciences is largely automated, and the epistemic circumstances are determined by computers rather than humans.[9] Often data-intensive science is used synonymously with data-driven science, which focuses on data analysis rather than complex theoretical constructs. This contrast is often expressed as deductivism versus inductivism, the former signifying analysis that starts with concepts and the latter signifying analysis that ends with concepts. Contemporary datasciences present themselves as strongly inductivist and often rely on a small number of general modelling assumptions, while conventional deductivist analysis is model- or hypothesis-driven. The field of datasciences mostly names the contemporary practice of applying machine learning algorithms to big data and develops epistemological and conceptual imaginary as a foundation for a strongly empiricist and inductivist approach in scientific and practical fields such as finance, insurance, healthcare, marketing and advertising. The underlying logic of datasciences is that, if social action unfolds in multidimensional distributions of socially efficient properties, which stipulate a set of patterned positions, analysis can predict strategies.

Although a shift from deductivism to inductivism is often seen as the foundation of datasciences, Wolfgang Pietsch (2022) is among the first scholars to advocate epistemological consequences of datasciences as a reinvention of inductivism. Like other scholars he defines the datasciences as the extraction of knowledge from large and complex data sets. By distinguishing between data-driven and data-intensive science, he highlights the increasing reliance on data and the development of complex models and theories in data-intensive analysis. He argues that the impact of the internet and the massive availability and deployment of sensors on the growth of data are accelerating this inductivism. Pietsch argues that artificial intelligence and machine learning in the datasciences are exponentially increasing the ability to analyse huge data sets and make predictions. The ten theses that Pietsch presents on datasciences, covering topics such as predictability, modelling, statistics and the changing role of scientific experts, captures the significant transformation we want to highlight between regulative power and sensory power. His central argument is that datasciences rely primarily on inductive inference, and that inductivism is a viable and reliable approach to both scientific and practical research. For Pietsch the emergence of inductivist paradigms in statistics is blurring the boundaries between supervised and unsupervised learning in machine learning algorithms. The challenges of cluster analysis and artificial neural networks in datasciences highlight the importance of unsupervised learning in developing effective concepts for prediction and influence. The significance of new inductivism is that it represents a shift from a type of analysis from statistics with its predictive modelling to another type of analysis where assumptions about the data are minimised or even eliminated. What we want to offer by tracing the development of concepts in the population game from surveillance, classification and correlation to correspondence is to illustrate the relations between these developments and the transformations from sovereign and disciplinary power to regulative power, and the associated forms knowledge, practices and methods. What we want to trace is a shift in the inventions of sovereign power, disciplinary power and regulative power that makes possible sensory power and how this shift precipitated a shift from a deductivism to inductivism. To put it differently, accounts given as internal developments in a form of knowledge can be challenged by their external relations to forms of power.

Surveillance

If we examine the emergence of concepts relating to data, surveillance is probably the oldest and closest to sovereignty. Although it has been used since the nineteenth century to denote the watching or guarding of a person, especially a suspect, a prisoner or the like, and is often associated with spying or surveillance, it is worth recognising a broader function of the concept as supervision for the purposes of direction, control or superintendence.[10] As David Lyon (2018) illustrates, watching and supervising people and things is one of the oldest strategies of power, producing both affects and effects, and inspiring the invention of numerous technologies of power. Yet, Lyon argues, the growth in sensing and storage technology, together with the increase in available data types such as text, images and video, led to the creation of large-volume, high-dimensional datasets that are used for surveillance purposes where surveillance changes its meaning. These methods reconfigure surveillance as they identify and make up different peoples, influence their conduct and perform supervision along with the meaning and function of punishment, which were associated with the invention of disciplinary power. Thus, the new surveillance is not only about watching, supervising and controlling peoples but also interpellating people into imitative practices where people react to being watched and become watchers themselves. The emergence of technologies of power centred on the body, leading to the spatial distribution of individual bodies; their separation, alignment, serialisation and surveillance were the products of disciplinary power using resignifying and repurposing surveillance as an adaptation to mimetic learning. The general system of surveillance and penal repression was set up to make previously tolerated illegality impossible. The development of the police apparatus and the surveillance of the population paved the way for the emergence of a new form of power-knowledge, which would take the form of an examination.[11]

As much as surveillance has been essential for exercising sovereign power, its transformation and adaptation have been vital for the invention of disciplinary power. Surveillance may not be immediately associated with doing things with numbers but arguably for both sovereign and disciplinary power surveillance prompted numerous data practices such as surveys and censuses, which were the original practices of the collection, gathering and interpreting data about territory, population and subjects. Yet, if these practices enabled the invention of the population game by regulative power that adjusted bodies to the demands of the population, surveillance has changed its meaning and function dramatically.[12]

Classification

The concept of classification has been important especially for disciplinary power and regulative power. If in the seventeenth century statistics was already used as an outline for describing the conditions of a state and its population, complex forms of classification facilitated to tabulate and present them for purposes of governing peoples through different kinds of surveillance. Doing things with numbers would have been impossible if they did not produce classifications of different people (children, adults, soldiers, paupers) and things (territory, population, wealth). Classification enabled doing things with numbers that they generated. So, before it invented metrics (as quantification), statistics was a practice of doings things with numbers (classification) – a double meaning that it retains still today. As Jon Agar (2003) observes, the collection, collation and tabulation of statistics has created a great bureaucratic machinery of classification, which often then thought of itself as providing only information but is itself part of the technology of power in a modern state. And, as Hacking (1990) shows, many of the modern classification and categories by which we think about people and their activities were put in place by doing things with numbers: counting produces categories, and many of the categories used to describe people are byproducts of the practices of enumeration (Hacking 1983, 280). Or, as Pierre Bourdieu ([2015] 2018) would say, the struggle, which is the very principle of the distributions of different forms of capital, is inextricably a struggle to appropriate rare goods and a struggle to impose the legitimate way of perceiving the power relations manifested by the distributions, a representation that, through its own effects, can help to perpetuate or subvert these power relations. For Bourdieu, classifications, including the very notion of social status, would not be such important objects of struggle if they did not contribute to the existence of the social classes, enhancing the objective mechanisms that determine the distributions and ensure their reproduction, and indeed by adding to them the consent of the ideas which they structure. The challenge is to move outside and place the modern polity in a general history of strategies and technologies of power that assured its mutations, development and functioning.

Classification has been so essential for organising practices and methods in doing things with power that, as Hacking (1990, 3) notes, it generated the idea that authorities can improve and control a deviant group within a population by enumeration. As Alain Desrosières (1991) shows, the concept of classification was the origins of the nineteenth-century problem of the correlation between the part and whole and the direction of their relations. This was the moment of the invention of regulative power, with its focus on the relation between bodies and population, and the direction that the population could provide for exercising power on bodies. The operative principle of regulative power calibrating the part and the whole made classification possible but classification would not have been possible without the invention of disciplinary power and its drive for surveillance as supervision.

How classification as a concept has undergone an important transformation in the twenty-first century is the most important development. The increasing volume and complexity of data (and the invention of methods capable of analysing such data) meant that extracting the relationships embodied in such data became the operative principle of analysis, rather than inventing categories to fit people into as regulative power demands. The logic has changed. Classification to calibrate the part to adjust or adapt to the whole (regulative power) became a recursive activity of discovering latent relationships as clusters in a given field of sensory power. This also means a fundamental change in the population game, as classification as latent relations has practically abandoned population as an object of government, thus rendering population as such less relevant to power. This does not mean that all logics of classification have become irrelevant; on the contrary, for sovereign, disciplinary and regulative power, the logic of classification continues to govern people. Yet, and this is our argument, the transformed logic of classification signals the birth of sensory power, but this will become clear only when it is linked to concepts of correlation and correspondence and eventually to learning machines.

Correlation

The discovery of the concept of correlation by Francis Galton in the 1880s is considered one of the greatest events in the history of science (Bulmer

2003). It also coincides with one of the greatest inventions in the history of government: regulative power. Galton's work on correlation and regression led to the development of biometric statistics, which focused on studying the variance of distributions. It became the dominant statistical technique in the social sciences. It was also at this moment that psychometrics emerged, providing statistical tools for differential psychology and the study of human intelligence, the importance of which fully appears in the imitation game. The concept of classification that served so effectively for disciplinary power for training and educating bodies, maximising their capacities and increasing their efficiencies operated with a kind of determinism that established relations between the needs of knowledge and the demands of power. As Hacking (1990) illustrates, the decline of determinism coincided with regulative power and its demands for taming of chance.[13]

To put it differently, it was the use of statistical laws to explain phenomena that were produced by classifications that led to what Hacking calls the taming of chance. Hacking says that the concept of laws of a population, once considered a conceptual impossibility, became routine by the nineteenth century. The drive towards analogue computing has led to the principle of correlation replacing the principle of causation. As Desrosières (2014) shows, the work of Adolphe Quetelet (1796–1874) extended the principle of least squares to include correlation and regression, and Charles Spearman (1863–1945) developed factor analysis, a statistical method used to show the interrelationships of mental abilities of different peoples. Such statistical work involves mathematical techniques that summarise categories into averages and correlations, revealing statistical constants and regularities in a sample that make it possible to estimate the population as a whole. If classification was both an effect and an instrument of disciplinary power, correlation was the principal concept of regulative power, finding ways of adapting bodies to the needs and demands of the population, finding patterns in parts to calibrate them with the whole, and finding ways of representing the latter in the former. The concept of correlation discovers that while there is an indeterminate or stochastic relationship between classes of people and the population they constitute, the art of governing this relationship or taming chance is the operative principle of regulative power. Both knowledge and power rely upon the assumption that hypotheses (knowledge) or directives (power) can be confirmed by the analysis of the whole and its parts. Governing the whole necessitates finding strategies and technologies of power that can adjust the parts to the desired whole. Both in knowledge and in power this operates as deductivism.

Yet, like the transformation of the logic of classification, correlation too has significantly changed for similar reasons: the accumulation of data and the invention of methods capable of analysing such data meant that correlation was becoming less concerned with adjusting the parts (bodies) to the whole (population), which was the logic of regulative power, and more concerned with developing techniques for identifying latent relations based on correlations in a given field of power.[14] That this transformation began before the development of digital computers in the 1940s and accelerated afterwards developing into the twenty-first-century datasciences should make sense especially with the development of the concept correspondence as opposed to causation.

Correspondence

It was in the late twentieth century that the concept of correlation enabled a new concept – correspondence – which became the operative principle of sensory power. The operative principle of modern datasciences is seen as inductivism, which means understanding any given cluster of relations without underlying theories, expectations or assumptions. This is the opposite of deductivism, where the taming of chance as an activity of government requires adjusting the parts to the whole, which was the logic of regulative power. We will discuss correspondence analysis later as a statistical method, where it is used to analyse complex patterns of relations, but here we will focus on it as a concept. Correspondence allows researchers to visually and geometrically construct the spatial and social distances between diverse variables and identify coherent sets of relations stemming from distinct systems of proximity. It has evolved from related methods such as principal components analysis, factor analysis and cluster analysis developed by Karl Pearson ([1892] 1911), Charles Spearman (1904) and Louis Leon Thurstone and Ernest John Chave (1929).[15] A key to understanding the logic of correspondence before it was named is Pearson's *The Grammar of Science* ([1892] 1911). As Desrosières ([1993] 1998, 108) notes, Pearson ([1892] 1911) makes a brilliant intervention in logic of science against causation. Desrosières says that this is a new approach against determinism but a closer inspection may reveal even something more radical.[16] Simply put, for Pearson reality is something that science produces and does not merely discover.[17] To put it in our words, the logic of science appears performative rather than propositional. It took about fifty years before this operative principle found its expression in analytical programming.

Correspondence analysis was first proposed in 1962 and has become a standard, unifying and integrated analysis of latent and previously unknown relations. It is often argued that correspondence analysis can only be used for displaying relationships (inductivism) and not for hypothesis testing (deductivism). The success of correspondence analysis is attributed to the principle of distributional equivalence and the availability of software applications. It is similar to cluster analysis but more efficient and quicker in identifying relationships between variables. Correspondence analysis does not aim to test hypotheses by providing levels of statistical significance, and other statistical techniques are required for testing hypotheses. If the technique has its roots in the factor analysis and cluster analysis developed by psychologists, it was transformed in the 1970s by the work of Jean-Paul Benzécri ([1973] 1992) and Brigitte Cordier-Escoffier (Desrosières 2016, 186). We will discuss below in the methods section the two-dimensional mapping of the social space based on correspondence analyses used by Bourdieu ([1979] 1984) and the importance of this for understanding the operative principle of sensory power. It is often said that correspondence analysis allows researchers to see their data and identify relationships between variables as a multifactorial form of statistics that reveals objects through their relations. The key importance is that by the late nineteenth and early twentieth centuries correlation and its associated methods, factor and cluster analysis, were already driven towards a conception of science diverging not only from determinism as Hacking noted, but a new relationship between the parts and the whole than how surveillance and classification played the population game (Desrosières 1991). To put it differently, the origins of sensory power can be seen in this shift from seeking relations between the parts and the whole (without normal and abnormal or the bell-shaped curve) to identifying latent relations in a given domain without making assumptions about any distribution. A new game was being invented by playing the population game, which will be named as the imitation game and will become the operative principle of machine learning in artificial intelligence.

These concepts in the invention of the population game should serve to illustrate the relations between knowledge and power and how concepts or their operative principles enable new forms of power and how new forms of power in turn produce demands for knowledge. From the late eighteenth century to the twenty-first century the developments of datasciences and its concepts such as surveillance, classification and correlation were intertwined with the shifts of sovereign power, disciplinary power and regulative power that made correspondence as an operative concept of sensory power possible. Doing things with numbers converging with doing things with neurons will make sense against this history.

Practices: survey, census, inquiry, mining

The internet and the increasing availability of sensors have made previously unthinkable modes of data collection possible, leading to a growing accumulation of data that is connected and connectable with other data. However, there is a growing restlessness about how all this data are captured and the assumption that underpin them. As Ruppert and Scheel (2021) demonstrate, data practices such as selecting, formatting, editing, storing, cleaning and analysis are iterative processes that produce data. Data practices are sociotechnical, situated, performed by actors as stakes in struggles over authority and power within specific professional fields of practice, contingent and enrolled in the enactment of their object. Data practices contribute to making up the very objects and subjects that they seek to represent. Or, as Agar puts it, 'an immense amount of physical and intellectual work is needed to create even the most simple statistical fact – indeed, the simpler a fact is, the more refinement and therefore the more effort, is necessary . . . and facts have to be manufactured, with all the implications of process, organization, quality, and product that the metaphor of the factory involves' (Agar 2003, 75). Agar's study of what he calls the mechanisation of government in the United Kingdom, focusing on the changing capacities of government from the early nineteenth century to the beginning of the twenty-first century, relates computing technologies and data practices. Agar shows how governing a state and the inventions of computing technologies and data practices became intertwined. He shows how the expert movements have become active since the nineteenth and twentieth centuries, including statisticians, economists, the 'Organisation and Methods' movement, and operational researchers. He examines the organisation behind the creation of statistical facts and the symbiotic relationship between official and non-official statisticians. The development of statistics and the interdependence of official and non-official sources were related movements. The establishment of the Statistical Society of London and the shift from enumeration to analysis in statistics was a turning point from statistics to analytics.[18]

Doing things with numbers means government by management and calculation characterised by orderly, objectifying and subjectifying procedures. They are central to persuasion because of the expectation that they will be accurate, precise and reliable. If doing things with numbers is historically contingent and culturally variable, and they can be used to do more than organise and convey information, then archologies and genealogies of data practice must form the background against which we can understand the historical relations between forms of knowledge and power. If words reproduce and reshape the cultural meanings of numbers for exercising power, then we must understand how data practices organised and are transformed. Data practices such as survey, census, inquiry and mining are each an invention by a specific form of power – respectively, sovereign, disciplinary, regulative and sensory – each having been transformed by the invention of a subsequent form of power.

Survey

The use of surveys for knowledge and power has a long history, with early surveys conducted by the state for purposes such as raising revenue or military recruitment.[19] Clearly, sovereign power always depended on reliable surveys of territories and subjects over which it ruled. Our interest is however on the emergence of disciplinary power and how it invents social survey as an affective and effective practice for not only gathering and analysing data but also producing subjects of data practices. This is the condition of possibility of the invention of the population game: that any given territory did not only comprise peoples but they cumulatively constituted a population with its specific laws. Adjusting and adapting people into these laws would lead to regulative power but not before disciplinary power inventing technologies for producing bodies.

The account given by Eileen Yeo (2003) shows the transformation of surveys from descriptive to constitutive surveys in the eighteenth and nineteenth centuries before the development of regulative power. Yeo traces the origins of social surveys to state-conducted inventories of property, provisions and people for the purpose of raising revenue or military force. Yeo then examines the shift in the purposes of studying populations that led to the transfer of surveys into the hands of other social groups. Yeo shows the emerging role of the state in social surveys, particularly in the context of the French Revolution and the move towards professionalisation by the beginning of the twentieth century. Of interest is the convergence of assumptions from peoples to population just before the nineteenth century. This also marks the moment of resistance to surveys and the national compilation of vital statistics in Britain, and the abandonment of statistical secrecy by most states after the French Revolution. These were indications that there were struggles over data. The influence of the modern French state on the choice of methodology for social surveys, as well as the focus on the working poor in Britain during the wars with France, may have shifted these struggles to another terrain. The proliferation of government and voluntary surveys on an unprecedented scale in the age of statistical enthusiasm made it impossible to distinguish official from unofficial statistics. Yeo argues that the focus on uprisings in large cities in France, Britain and the United States, as well as the early socialist movement's opposition to survey practices, may have rekindled an interest in variation and diversity of types, as well as identifying diversity of groups, as opposed to earlier attention to and a preoccupation with averages. This is where we see the decisive signal of the development of regulative power. The expertise of the mid-nineteenth century on the working classes, as well as the intense focus on physical efficiency and heredity by the late nineteenth century, and the role of female investigators in social surveys, particularly in Britain and America, point to what Foucault highlighted as the diffuse and distributed involvement of different actors in the activity of government in exercising the emerging regulative power.

By the end of the nineteenth century, surveys were also being used by other social groups for different purposes, such as studying populations to identify pathological and dangerous peoples to ostensibly ameliorate class struggles. In England, for example, working men formed friendly societies for self-insurance, which were regulated by a law and encouraged mutual support.[20] Although these societies were often too small to have broader effects, their affects and effects were considerable for establishing a strong relation between data accumulation and subject accumulation and establishing actuarial patterns in determining the statistical laws of wellness and sickness.

By the twentieth century, immense amounts of survey data were accumulated and analysed. One important development in the use of surveys was the shift towards making participation in research a civic duty, rather than relying on the network of sociability. This was exemplified by the English statistician Arthur Lyon Bowley's (1869–1957) survey of one in twenty of the population in four towns, chosen to represent both mono-industrial and multi-industrial areas (Desrosières [1993] 1998). Bowley recognised that conditions of comprehensiveness and representativeness were linked to the obligation to respond to a survey. A decisive moment was the use of surveys to compare the condition of the working classes in different countries, driven by the principle that the vitality of the population was crucial for national competitiveness. Methodological breakthroughs in social survey methods followed. The use of averages and the concept of the 'average man' in social research were abandoned, Adolph Bertillon arguing that the mathematically derived traits of the average man were rarely to be found in actual subjects or populations (Stigler 1999).

If the use of surveys has evolved over time, from state-conducted inventories to voluntary surveys focused on social pathology and class conflict, the development of statistical laws of population such as sickness and the use of surveys to compare the condition of the working classes in different countries were the practices of regulative power for the measurement of the mutual adaptation of peoples and populations. The present data practices of instantaneous, continuous and live surveys that lead to metrics and rates are not based on survey practices instigated by disciplinary power and regulative power but transform their operative principles to suit sensory power. And, as we will see, once the reaction to survey results is automated, surveys are transformed into signals and pulses.

Census

With so much interest in the census and power, we can now easily claim that census practices have been around for centuries, with the earliest surviving and well-known ancient census probably originating in ancient China.[21] Gunnar Thorvaldsen (2018) shows that it is possible to produce a comparative history of censuses and census-taking around the world, which indicates not only the similarity of practices but also their logics. The earliest recorded censuses may have been in China during the Han Dynasty, or the Roman census organised by Emperor Augustus. Yet for censuses as practices more sophisticated than surveys, crossing from disciplinary power to regulative power, we must turn to the history of colonial censuses in British America, particularly in Virginia and the objectives of these early population counts.[22] The importance of the censuses in assessing military strength and economic potential in the American colonies was immense and the challenges and opposition faced by census takers was considerable.[23] Thorvaldsen especially explores the role of censuses in colonial control as population and territorial control, particularly in the French colonies in Quebec and the Danish census-taking in Norway. While Thorvaldsen notes the resistance against census-taking in the metropoles of England and France, particularly by the aristocracies, he argues that the colonial censuses were ordered without serious opposition, driven by mercantilist interests and the will to govern colonies without disrupting metropoles. Thorvaldsen illustrates how census-taking in the eighteenth century, with Denmark organising reporting from parish ministers and Sweden setting up Table Works, led to the French demographers figuring out population sizes using vital statistics.[24]

The boundary between a survey and a census may appear thin but conceptually the frequency and persistence of the latter enables the invention of regulative power. Census-taking became more routine after European colonisations 500 years ago, with the British and French leading the way in both colonial and metropole censuses in the nineteenth century as an exercise in taking stock of their growing empires (Glass 1973; Higgs 2004). The most ambitious were the decennial censuses of India, starting in 1871 and undertaken in the name of efficiency and welfare reform (Yeo 2003, 92). It is important to recognise that, while there has been resistance to census-taking, particularly in England and France before the nineteenth century, in the colonies census-taking was ordered without serious opposition, as imperial powers wanted to see their colonies develop without hurting domestic growth by sending more emigrants overseas. The original drive behind census-taking in the colonies was to assess whether their military strength was enough for their survival, with the report in 1676 about 3,582 British men able to take up arms compared with 9,720 French, raising concern both in London and in the colonies (Thorvaldsen 2018, 16). Gradually, the census has been used as a practice for population control and surveillance, with population registers and joined-up government administrative databases being investigated or implemented as alternatives. Yet the census has also had unintended effects, such as subversively demanding kinds of things or people to count (Ruppert 2012). The census has been revolutionary in marking the beginning of the routine generation by the government of statistical information that was not of direct administrative utility, precisely the 'informative' statistics that were at issue. The Royal Statistical Society began to mention and encourage the use of punched-card machinery in advising the government in preparation for the 1901 census, and the 1890 census invented the punch card, the chief development of early computing devices (Agar 2003, 148; Hacking 1983, 290).

The census has always produced contestable and changeable knowledge. The British census regarded home-based women as productive workers in the mid-nineteenth century, but by 1881 had started to move them into an 'unoccupied' class of unproductive dependents, a designation that feminists all over the world were disputing at the turn of the twentieth century (Yeo 2003, 92). The practice of the census exemplifies what Bourdieu called classification struggles, to indicate that class struggles for power have always involved struggles over the classifications by which subjects or peoples find themselves in social space. Yet, in the twenty-first century, the census has become a costly practice, partly because of the struggles over its effects and implications but also because of questions raised in the twentieth century about whether it can provide the required data sufficiently quickly and at a sufficient level of localisation and detail, integrated with other data sources.[25] This was, as Ruppert and Scheel (2021) indicate, a significant transformation in the twenty-first century: the shift from the census as a state practice to a corporate practice of data capture and mining. This has of course given rise to an immense debate over privacy and anonymity and yet simultaneously masked another and perhaps most consequential aspect of it: the monopoly of power over the knowledge about subjects. The capture and mining of data from pervasive devices and sensors has not only ushered in the invention of sensory power but also resignified previous data practices invented by disciplinary power and regulative power. The debate about the census in the twenty-first century, whether to continue the practice at all, masks a struggle between regulative and sensory power, the hegemony of its respective actors, and the monopoly of authorities (Isin and Ruppert 2019).

Inquiry

The practice of inquiry is a technology of power that authenticates truth and acquires and transmits things that are regarded as true. It embodies a form of knowledge and power analysed most memorably by Foucault to understand the relations between conflicts over knowledge and disciplinary power. In the historical formation of measure and examination, inquiry has been both means of exercising power and rules for establishing knowledge. Foucault shows how measure establishes or restores order and disorder, while inquiry establishes or reconstructs facts, events, acts, properties and rights. These are also related to examination that fixes or restores the norm, rule, division, qualification and exclusion (Foucault 2019b, 221). These methods were bound up with the invention of disciplinary power as its effect and instrument and were developed by regulative power. The judicial model of inquiry rests on a whole system of power that defines what has to be constituted as knowledge, how it is extracted, transferred and conveyed, and at what point it accumulates and gives rise to a judgement or decision.[26] Foucault would explore how inquisitorial societies practice the extraction, transfer and accumulation of knowledge to enabling accumulation of subjects.

If Foucault famously identifies two stages in the invention of disciplinary power and then regulative power, the former targeting the body and the latter targeting population and its metrics, this raises the question of how population metrics were taken to govern the conduct of bodies. This led him to distinguish or at least allowing us to distinguish between power and government as distinct activities. Governing bodies after disciplinary power necessitated strategies and technologies that involved positive engagement of people in their own government. It was here that early forms of inquiry or inquisitional practices were transformed into pastoral technologies of government. It is almost as if the accumulation of subjects and accumulation of capital would have only been possible with the incorporation of people into their own government through pastoral technologies of power and inquiry transforming data practices into complex technologies of power for calibrating between bodies and populations through intermediary cluster of relations.

However, if in the late twentieth century, based on his studies of the period between the late eighteenth and nineteenth centuries, Foucault (2019b, 222) would claim that 'we belong to an inquisitorial civilisation which for centuries has practised the extraction, transfer and accumulation of knowledge in increasingly complex forms, all derived from the same mode', we must now ask whether data mining has come to define the extraction, transfer and accumulation of knowledge in the twenty-first century and the birth of sensory of power to which it is linked. To put it differently, can we read mining as a form of inquiry driven by and enabling sensory power?

Mining

If the mutual adjustment between bodies and populations did not require inquiry, how would this change the government of peoples? As regulative power enables a new form of power – sensory power – we recognise that bodies are becoming clustered to govern themselves as self-regulating, self-organising and self-controlling members of these clusters. When numerous dynamic and instantaneous data flows through the internet to feed the sensory data of movements and actions, signs of bodies, it creates more data than can be effectively analysed. These sensory data are ephemeral in nature and cannot be stored for too long, posing a major challenge to collation and tabulation systems. What can be done? If we mine these sensory data by classifying and associating, reducing their complexity into patterns and shapes, collating and tabulating them and storing only these results, we will require new methods.[27] Following factor analysis and anticipating correspondence analysis, cluster analysis is the name for this method that is both the effect and instrument of the demands of datasciences in the twenty-first century. If we consider several applications where clustering has been used, such as image segmentation, object and character recognition and document retrieval, data mining becomes a dynamic and continuous process. This is because data mining is a continuous data practice that can be applied to relational, transactional and spatial datastores, as well as large stores of unstructured data collected and tabulated from the internet.[28] This is where machine learning arguably becomes learning machine, a data practice that reconfigures local centres of power and knowledge by redrawing human–machine relations, and a major shift in the history of statistics as the science of learning from data.[29]

We have only hinted at the emergence of new demands for knowledge and power, and how these led to the invention of sensory power and mining practices as a form of knowledge to meet these demands. We will now turn our attention to more specific methods, especially since the seventeenth century, involving the invention of statistics, analytics, clustering and learning as ways of doing things with numbers, which both stimulated and responded to different forms of power and enabled the practices of mining which we have just identified as the inheritor of ancient practices of inquiry.

Methods: statistics, analytics, clustering, learning

This much is certainly agreed: the increasing use of social networks, platforms and the internet, together with advances in computing power and data storage, have led to the production of massive amounts of new data for analysis.[30] This has apparently led to the emergence of data-driven science, which adheres to inductivism and data objectivity.[31] This view signals that there are struggles over the meaning of the methods that are appropriate for analysing these massive amounts of data.[32] We now see data assemblages and data infrastructures as complex and integrated systems, involving both technical and cultural arrangements.[33] If in the nineteenth and twentieth centuries the production of statistical facts required immense physical and intellectual labour, and statisticians formed a professionalising movement to influence government through the production of statistical facts, in the twenty-first century the struggle is over the design, deployment and maintenance of data assemblages and data infrastructures by a new class of professionals, datascientists, technologists and entrepreneurs. If in the nineteenth and twentieth centuries the establishment of central statistical offices and the routine production of statistical information not of immediate administrative use marked the beginning of informative statistics (rates, indexes, scales, metrics), has the twenty-first century ushered in performative statistics? If the campaign for an imperial statistical office revealed tensions within the professional movement of statisticians in the nineteenth and twentieth centuries, do the struggles over platforms among data empires reveal the emergence of new technological classes? If doing things with numbers has always depended on establishing normative legitimacy through accurate and precise informative statistics, what are the operative methods of datasciences in the twenty-first century that establish normative legitimacy through performative statistics?

There are implicit answers to these questions in the discussion above of how concepts such as surveillance, classification, correlation and correspondence developed intricately together, and how they stimulated and were influenced by data practices such as survey, census, inquiry and mining. Interwoven with these developments were methods of datasciences: statistics branched out into analytics, clustering techniques were developed to respond to the needs of analytics and learning from immense amounts of data evolved into machine learning methods, leading to claims of machine intelligence and intelligent agents. We cannot, however, provide full answers until we recursively discuss specific methods of datasciences against concepts and practices that have generated demands from different forms of knowledge and power, and how heterogeneous elements were assembled to generate these methods. In other words, it is impossible to understand methods such as statistics, analytics, clustering and learning without understanding them within an ensemble of concepts, practices and methods arising from different but evolving forms of knowledge constituting datasciences and their associate forms of power: sovereign, disciplinary, regulative and sensory.

Statistics

The function of statistics in shaping knowledge and power in polities has become a central theme in the history of statistics since the 1980s. The challenge of distinguishing the social reality produced by the social sciences from that produced by social forces, and the emergence of statistical reasoning as a method of both totalisation and individualisation, as well as the function of statistics in moving from the parts (bodies or peoples) to the whole (populations), has generated an immense body of scholarship as we have discussed above. The history of statistics as a social and political activity, focusing on the emergence of scientific and non-scientific societies and official and non-official statistics (and statisticians), has been studied with some intensity. This scholarship has also linked all these developments in concepts and practices to specific methods and their history.[34] The transformation of scientific principles for describing the social world and the emergence of the modern idea and technical deployment of representativeness has been shown to be a social and political event.[35] The importance of providing a history of objects and subjects and how they come together in statistics has been documented, as well as the double historical origin of statistics as political-administrative and social-scientific practices embodied in the word 'statistics'. The role of statistics in creating laws about society and the character of social facts, and in creating concepts and classifications within the human sciences, is revealed. The use of statistical laws to produce and explain biological facts and to fit them into heredity with the role of statistics in establishing a 'calculus of felicity' and the development of the social sciences.[36] The development of statistics as a form of knowledge was driven by the desire to govern populations, with statistics being used to classify, enumerate and regulate peoples, whose habits determined the laws of population expressed in vital statistics. We have discussed some of this scholarship above but here we will highlight how methods known as statistics enabled the advancement of analytics.

Although the history of statistics dates back to the seventeenth century, with the origins of probability theory found in the correspondence between Blaise Pascal (1623–62) and Pierre de Fermat (1607–65), it is significant that the theory of probability did not emerge until much later, for various reasons including religious and moral objections and the absence of regulative power as a force or drive. But the 'normal' and 'deviant' names in statistics emerged in the nineteenth century based on probability theory, which relates exactly to the political problem of adjusting bodies to population or the parts to the whole we have seen earlier. In the late nineteenth and early twentieth centuries, statisticians such as Francis Ysidro Edgeworth (1845–1926), Karl Pearson (1857–1936) and Charles Spearman developed methods for statistical analysis, including the study of evolution and the concept of general intelligence. Pearson formed a biometric laboratory and a eugenic one at University College London, and with Walter Frank Raphael Weldon (1860–1906) he formed the journal *Biometrika* for the study of life through calculation and measurement. The representative sampling was promoted by opponents of statistics, but probabilistic sampling became popular among social scientists, with Arthur Lyon Bowley defending random sampling and introducing its mathematics. The philosopher and mathematician Gottfried Wilhelm Leibniz (1646–1716) had argued for the establishment of an information bureau in Prussia to assemble vital statistics and census figures. Adolphe Quetelet gave body to the mean with his construct of the average man as Bowley developed techniques of random sampling and used the mathematics of probability and standard deviation tests to calculate the margin of error. Jerzy Neyman (1894–1981) settled the matter of probabilistic sampling with his analytical rigour.[37]

The new scholarship has shown not only that the growth of statistics was made possible by the invention of distinct methods based on mathematical probability but also that it created sociological and psychological phenomena. That order is to be found in large numbers became a key doctrine in nineteenth-century statistical thinking, with the regularity of crime, suicide and marriage being used to justify the application of statistical methods to problems in biology, physics and economics and extending to psychology and sociology. The elaboration of this view led to the erosion of determinism and has profoundly influenced the scientific world view of the twentieth century and the relations between the parts (individuals and bodies) and the whole (population). The emergence of probability and its development into stochastic analysis coincides with the modern population game.

If the invention of statistical techniques such as probability became historical and social events because they gave rise to social disciplines such as psychology and sociology then how could we account for the transition from probability to stochastics, from prediction to intervention? With the idea of probability replacing the concept of deterministic human nature, the transition from the seventeenth and eighteenth to the nineteenth and twentieth centuries was complete. That this transition of statistics from the seventeenth and eighteenth centuries to the nineteenth and twentieth parallels Foucault's argument about the emergence of two different forms of power – disciplinary and regulative – is not accidental. Although not a single source of inspiration, Foucault's influence on this scholarship is undeniable and it is inevitable that he has provided a perspective from which to view the invention and transitions of statistics in this period. However, a strong argument can also be made that this scholarship has filled many gaps in Foucault's investigations and has confirmed, refined and revised his conclusions, particularly his view of the two-stage development of regulative power. In any case, a reading of this scholarship without this perspective on power would be inadequate. Moreover, if we are now to ask about the twentieth- and twenty-first-century trajectories of statistics, such as analytics, clustering and learning and their relations to the birth of a new form of power – sensory power – this scholarship is absolutely necessary yet insufficient. For, while it has been brilliant, it has also remained within Foucault's scheme (with differences between scholars) and too insistent on fitting the histories they have provided back into Foucault's scheme of sovereign, disciplinary and regulative power. Writing mostly in the 1980s and 1990s, what we think they overlooked is that a barely sensible or intelligible event was happening that was not about statistics as such but was nevertheless influential in the articulation of sensory power: the convergence of datasciences and neurosciences since the 1930s. This is one of the aims of our reworking of power.

While the new quantitative science of statistics developed in the nineteenth century with a focus on addressing questions of disease, poverty and crime, eschewing determinism in favour of stochastics, probability was used as a model or theory to structure the physical and psychological characteristics of individuals and populations (and as Foucault would insist mutually adjust and adapt them to each other), with the combination of circumstances accounting for the physical features of men and women, as well as moral traits. That Francis Galton (1822–1911) became interested in the tails of the error distribution, where exceptional individuals were gathered, and became effectively the founder of eugenics was not an aberration but a logical consequence of the will to regulative power to mutually adjust and adapt individuals and populations.[38]

All forms of power accumulate costs that their exercise cannot eventually bear. The regulative power demanded too much and perhaps achieved too much. The mutual adjustment between individuals and populations required building an immense infrastructure of institutions and created complex organisations such as militaries and corporations that made the exercise of regulative power increasingly costly. Probability theory was the operative logic of regulative power but by the early twentieth century it was already being transformed into stochastic and heuristic methods that were to result in machine learning. Two such methods – principal components analysis and factor analysis – were originated by Pearson, while cluster analysis was initially proposed as an inductive method for analysing data. Associated with these methods were psychometrics and multidimensional or multivariate data analysis as it has now become, namely the 'search by induction of the hidden dimensions that are defined by combinations of primary measures' (Murtagh 2018, 148; 2019a, 25). Although made possible by the theory of probability, these methods were responding to a problem that simultaneously political and technical and began shedding the name statistics and presented themselves as analytics. Therefore, our focus in what follows is on how the problems of doing things with regulative power forced the generation of new forms of knowledge that in turn prepared the conditions of possibility of a new form of power – sensory power.

Analytics

The difference between statistics and analytics involves the distinction between probabilistic and stochastic processes, determinism and indeterminism and data-intensive and data-driven approaches to collecting, tabulating and analysing data. These differences also relate to a form of power that drives the need for collection, tabulation and analysis. As Hacking (1990) argues, a significant change in the late nineteenth century was from probabilistic to stochastic methods, as there were new demands to understand people without considering them as representative samples of a population or needing to predict their habits. Rather, the focus shifted from understanding peoples (the parts) that constitute a population (the whole) to making up people without imposing or direction by the population's requirements on their degrees of freedom. As Hacking would argue, this did not mean that constraints disappeared but their logic and operation changed. There is of course a very complicated history here that we are simplifying. How the population game that eventually made possible the imitation game that is our focus.

There is a remarkable resonance between analytics, which became increasingly sophisticated in the twentieth century, and the changing logics and forms of government of peoples. This involved the transformation of statistics into analytics, whose techniques were based on statistics but transformed to produce different kinds of methods for different kinds of problems. The emergence of multivariate statistics such as principal component analysis, factor analysis and cluster analysis was such a response. We can think of analytics as a period of transition from statistics (as the relation between the parts and whole) to the exploration of different ways of dynamically analysing data without making too many or any assumptions about its structural patterns for understanding the parts that form diverse wholes.

That these methods later coalesced into geometric data analysis (an approach to multivariate statistics that emphasises the geometric structure of data used for analysing very large and complex data sets and identifying patterns) may not be apparent but it indicates that the population game was transforming. It is this transformation that led to deploying supervised and unsupervised learning in machine learning algorithms and indeed the language of supervised and non-supervised classification or labelled and non-labelled data was first used in factor analysis and cluster analysis. Supervised learning involves training a model on labelled data, while unsupervised learning involves finding patterns in unlabelled data. Both approaches have their own strengths and weaknesses and are used in various applications, such as image recognition and natural language processing. The historical significance of statistics as the science of learning from data became more apparent when the development of statistical methods in the late nineteenth century and their application in various fields, including eugenics, biometrics, psychometrics and social sciences, led to the development of analytics. The struggles over legitimate methods for producing unofficial statistics and official statistics intensified transnational struggles over methods as analytics increasingly became agnostic towards statistics.[39] The challenges of standardising statistical methods across different countries and cultures and the importance of transparency and accountability in statistical reporting also resulted in the increasing supremacy of analytics. The reassembling strategy in establishing the superiority of established statistical methods into analytics was arguably giving birth to a new game – the imitation game. The approach that involves breaking down complex statistical models into simpler components and testing each component individually to ensure that the model is accurate and reliable will prove immensely useful in making brain–machine analogies. If the significance of Galton's discovery of correlation in the history of science led to the development of techniques for creating equivalence between different populations, including the method of random sampling, analytics became increasingly focused on learning from data without assumptions that appeared to be how cognition functions.

The various types of reinforcement learning, combining supervised and non-supervised methods of learning, including exploration versus exploitation emerged as ways of applying these methods. What is called reinforcement learning is for example a type of machine learning that involves training a model to make decisions based on feedback from its findings. Such methods are now applied in fields as diverse as medication dosing, portfolio management and drone navigation. These analytics demonstrate the versatility and deployment of reinforcement learning in various fields. Before we examine the transformations of these learning methods, we need to briefly discuss cluster analysis that underly these methods. Again, we are condensing an immense history here but the point is to see that the languages of supervised and unsupervised learning, reinforcement, labelling and learning are interwoven with the demands of a changing population game where population no longer means an already conceived body (as a whole) to which bodies (parts) must be adjusted but the operative principle of making up collectives that are neither parts nor whole but nonetheless constitutes objects of government – clusters.

Clustering

We have mentioned that the transition from statistics to analytics in the emergence of datasciences involved methods called clustering. Although clustering as a method of organising data into groups based on their intrinsic characteristics or similarity has a longer trajectory, its transformation into a named analysis – cluster analysis – in the 1930 and 1940s also marks the beginnings of machine learning.[40] It emerged as an exploratory and heuristic technique that aims to find structure in data rather than confirming an assumption about a population.[41] The distinction between confirmatory and exploratory analysis is significant as the former signals conventional statistics to confirm representativeness of data against a population and in the latter the aim is to identify the latent relationship within data.

Clustering has an intriguing history in various scientific and practical fields, and numerous clustering algorithms have emerged. There is no single clustering algorithm that dominates all practical fields. Clustering algorithms are broadly divided into two groups: hierarchical and partitional (Murtagh 2016). If the absence of categories distinguishes data clustering (unsupervised learning) from classification or discriminant analysis (supervised learning), the development of clustering methods has been an interdisciplinary movement across various fields. Originally, the clustering problem proved difficult, and differences in assumptions and different fields of application and have impeded the development of generic concepts and methods.[42] Yet this exactly proved its strength in the emergence of datasciences. Clustering has many applications, including image segmentation, object recognition, document retrieval and data mining. Since there are numerous clustering algorithms, the results must be interpreted and evaluated based on the field of application and the data analysed.[43] The analyst must have an adequate understanding of the specific algorithm used and the details of the data collection process. The similarity between the objective functions of clustering algorithms can be estimated by the similarity of the clustering solutions they produce.

The term cluster analysis first appeared in 1939, superseding synonyms such as Q-analysis, clumping, taxonomy and typology.[44] The Classification Society was founded in 1964 to share concerns and approaches to data analysis across disciplines. Karl Pearson is credited with originating principal component analysis and related methods of factor analysis. If we recall that deduction and induction are two methods of deriving general principles from particular consequences, or from knowledge of particular facts to general concepts, cluster analysis emerges as a prime mover of the inductive approach. When cluster analysis is used as an unsupervised learning technique that groups data points into distinct categories based on their similarities, it also generates new problems. It certainly solves the problem of recognising patterns without assumptions but since there are now some thirty-five algorithms to cluster data it can generate unreliable clusters. It can also be costly to apply it to immense amounts of data only to end up with meaningless clusters (Jain et al. 2004, 658).

The drive to mitigate against these problems by first confirming that there are meaningful clusters in a data set before computation of expensive cluster analysis and second to select the appropriate algorithm for the type of clusters has generated novel methods in datasciences. The concept of ultrametricity, for example, has been proposed to assess the clusterability of a data set (Simovici and Hua 2019). Similarly, semi-supervised learning combining supervised and unsupervised learning, and reinforcement learning methods have been introduced to determine the best sequence of actions to take over time to maximise some objective. If cluster analysis uses the similarities and differences between data points and data cases to uncover relations or hierarchical structures in data and if the choice of clustering algorithm depends on the objectives and expectations about the geometry of the clusters, this raises the question of whether cluster analysis is merely inductive or whether it is performative. When clustering first appeared it also referred to unsupervised learning or segmentation and it was used in psychology and commerce. The mechanical application of clustering algorithms does not always result in a meaningful partitioning of data. Therefore, the analysts must have a thorough understanding of the particular algorithm being selected and the details of the data collection process to ensure that the output of a clustering algorithm is evaluated based on the data domain. This means that the analysts must have a sense of what they want to do with numbers.

The word algorithm was used in the early nineteenth century in mathematics to denote a procedure or set of rules used in calculation and problem-solving. With the development of computing, it expanded to denote a precisely defined set of mathematical or logical operations for the performance of a particular task. As we mentioned, clustering methods use about thirty-five different algorithms. It is clear that data collection and representation are critical factors in the performance of clustering algorithms and the importance of selecting appropriate features for a specific field of application is important. If indeed clustering algorithms identify clusters in data whether they are present or not, and the need for different approaches to determine an appropriate algorithm for a clustering task emerges it is difficult to call any analytics based on clustering methods as 'inductivist'. Moreover, contrasting deductivism and inductivism as opposite epistemic logics is problematic. The evaluation of clustering algorithms demonstrates that the performance of a clustering algorithm reflects the data domain. The need to focus on the differences associated with machine learning applications in the production of knowledge becomes obvious. The challenge of machine learning applications is demonstrated by the assumptions about similarity and proximity and the data infrastructure that they represent illustrate that they are neither deductive nor inductive but performative.

Learning

When the invention of factor analysis led to cluster analysis in the 1930s, two distinct but overlapping developments followed. The first led to the emergence of correspondence analysis in social sciences. The second led to machine learning, used in speech recognition, pattern recognition and language models. That these two developments originate from the same methods of cluster analysis does not mean that they are deployed for the same objectives but both their similarities and differences are revealing. Their shared origins also raise significant issues for the uses of quantitative methods in social sciences, which Alain Desrosières (2016), Pierre Bourdieu ([2015] 2018) and Luc Boltanski (2016) emphasise when there is neither deductive nor inductive data analysis but always and only performative analysis in the Austinian sense that we use in this book.

We discuss the invention of correspondence analysis in social sciences first not because it happened earlier but because it highlights significant developments in the quantification of social science parallel to the invention of the imitation game. Developed as a relational method to processes large data sets, correspondence analysis is used for inductive reasoning and synthesis. Although it is seen as an integrated analytical method that encompasses statistical modelling, predictive analytics and visualisation, it is mostly used as a robust analytical method of visualising a social space of relations without making prior assumption about that space. Correspondence analysis not only identifies relationships between variables more efficiently than cluster analysis but also enables unsupervised classification. This means that unlike factor analysis it does not have to categorise (or label) predetermined variables (or dimensions) in order to create a space of relations. To put it another way, unlike factor analysis, it enables identifying complexity rather than reducing it.

That correspondence analysis had a parallel but overlapping trajectories in practical applications and social sciences is illuminating for understanding its uses. We have seen that cluster analysis too enables creating a space of relations often based on a transformed similarity or proximity matrix used to reduce complexity among known variables (or dimensions) of a dataset. By contrast, correspondence analysis attempts to establish a space of relations without reducing its variables (or dimensions) but arrives iteratively at a space of intensive and extensive relations. In a very strong sense, and this is the key point here, correspondence analysis learns from the data that it analyses and it involves unsupervised learning meaning prior assumptions and hypotheses are not made about the data itself. In this sense it is also neither inductive nor deductive but performative: its learning requires the intervention of the analyst appropriate to the objectives of the analysis. Thus, performativity of correspondence analysis is not any more or less robust than objectivity of inductivist analysis or any more or less robust than subjectivity of deductivist analysis. There are assumptions made about how, where and why data are collected and constructed, and the social conditions of possibility of its existence in the first place.

It now appears obvious why Pierre Bourdieu ([1979] 1984; [1984] 1988) was a prominent user. He used it as a method to show social spaces of relationships especially in demonstrating how social classes are distributed in it. The shift from statistics to modelling in data analysis and the use of visualisation to powerful description involved Bourdieu's adaptation of Jean-Paul Benzécri's ([1973] 1992) techniques of multiple correspondence analysis to operationalise the notion of social space. Bourdieu was aware that correspondence analysis was performative: he never claimed to use correspondence analysis for explanation or prediction. Although statistical methods before factor analysis and cluster analysis, such as multivariate analyses are used for explanation and prediction, correspondence analysis, which is also called geometric data analysis, are not recommended for using hypothesis testing. The correspondence analysis, or geometric data analysis, provides researchers with methods for understanding complex relationships between variables in social science research. Bourdieu used correspondence analysis or geometric data analysis for determining and representing the structure of social fields, showing structural homologies between fields, determining relative autonomy between fields, studying subspaces in a social space, explaining social practice and position-taking, and studying field dynamics using hierarchical classification.[45]

The two contradictory approaches to correspondence analysis illustrate this tension. If researchers approach analysis with a statistics episteme seeking to discover representativeness, they tend to follow a modelling approach with explanation and prediction. If researchers approach analysis with an analytics episteme, they tend to follow a learning approach without explanation and prediction. Bourdieu clearly approached it with analytics episteme. Bourdieu expressed serious doubts about a statistics episteme and used correspondence analysis for mapping social spaces of relations. Moreover, as Desrosières (2016) argues, correspondence analysis if used to make claims to explanation and prediction serves the purposes of regulative power and warns social scientists against such uses. Fionn Murtagh (2019b) goes further and argues that Bourdieu's work particularly in relation to social space and conduct constitutes the key developments in machine learning in datasciences. Murtagh says that the methodology of determining and representing the structure of social fields, showing structural homologies between fields, determining relative autonomy between fields, examining subspaces in a general social space, and studying field dynamics using hierarchical classification are the foundations of modern datasciences. For Murtagh, methods such as statistical modelling, machine learning, predictive analytics and visualisation analytics are related to correspondence analysis. Although he insists that representativeness and assessment of external validity are not the objectives, he nevertheless believes that the distance between 'data analysis' and 'data analytics' can be bridged through the use of multiple correspondence analysis. His conclusion that Bourdieu's work is central to datasciences, and the developing field of machine learning are contentious claims as we shall see below.[46]

The claims that Bourdieu uses correspondence analysis for explanation are also made. Cockerham and Hinot (2009) use correspondence analysis to quantify habitus against such claims. The concept of habitus refers to enduring dispositions towards action in social situations and settings. Habitus incorporates external social structures and individual inclinations, preferences and interpretations. Cockerham and Hinot propose the use of multilevel modelling techniques like hierarchical linear modelling to measure the effects of habitus on social behaviour. They argue that correspondence analysis, while efficient in identifying relationships between variables, cannot be used to explain or predict.[47]

Luc Boltanski (2016) reminds us of the question posed by Desrosières: 'how we can obtain knowledge about the social world (or produce statements that meet a requirement for truth about the social world) when all the instruments available to us to do so are given to us by the social actors whose relations with each other constitute that same social world and, moreover, it is only on this condition that they can move towards a meaning' (Boltanski 2016, 36)?[48] Since in contemporary societies power is largely exercised by doing things with numbers, Boltanski argues, 'this means moving from numbers that claim absolute validity and are therefore likely to reinforce absolute power to numbers that are in constant dialogue with other numbers, that is, if you like, to reflexive numbers' (Boltanski 2016, 42).

Bodies, populations and clusters

If doing things with numbers is always performative rather than deductive or inductive, reflecting data infrastructures and algorithmic structures, we cannot understand the formation of polities as clusters as either objective or subjective processes. This has been the case over the last two centuries with statistics and analytics and it is not otherwise with clustering and learning. Whether assembled through surveys, censuses, inquiries or mining, the quantitative data has always outstripped what can be done with it. If this led Hacking to claim an 'avalanche of numbers' between 1830 and 1848, he also observes something qualitative. He recognises that the overt reasons for doing things with numbers have to be distinguished from what he calls doing subversive things with numbers, which were not subversive in the sense that we might use the term today, meaning the thwarting of a technology of power, but subversive in the sense that Austin (1962) uses: infelicitous. As we have seen, Austin insists that when people are doing things with words they rarely control whether the words they use will have the effects or affects they intend. We can think the same about doing things with numbers. Hacking shows that many of the categories by which we govern ourselves and others were invented in the process of assembling and analysing data, whether or not they served the purposes for which they were intended. Today we can fear that we are drowning in numbers, but the challenge is to reveal the subversive or infelicitous uses for which numbers function and their affects and effects.

We now see the outstanding scholarship of the last thirty or forty years on the history of probability and statistics, understood as both information and analysis, from a different perspective. We now see the concepts, practices and methods of datasciences not only against sovereign, disciplinary and especially regulative power but also anticipating the birth of a new form of power – sensory power. If we have developed a sufficiently robust perspective on doing things with numbers and its operative concepts, practices and methods to focus in particular on sensory power and its directives for specific a form of knowledge now named as 'datasciences', this will enable us to understand the formation of neurosciences.

We have identified four operative concepts that organise different forms of power: surveillance, classification, correlation and correspondence. The boundaries between these concepts and their mapping onto different forms of power are blurred, but clear enough to sharpen the stakes of their transformations. Surveillance is a relatively straightforward technology of power, producing forms of knowledge suitable for observing and monitoring people and things. Classification is more sophisticated, producing not only categories of people but also directives to govern their behaviour. Correlation makes it possible to establish a relationship not only between desired conduct and bodies but also between the desired population as a rich and healthy entity, in order to shape and adjust bodies mutually. Correspondence is about forming clusters of bodies and things as relationships and, without making assumptions about the parts and the whole, enabling these clusters to govern themselves as self-organising collectives.

The order in which we have discussed these operational concepts is arbitrary. Echoing Hacking, we might say that there is a dynamic loop between them. There is no sequence. But there is accumulation: surveillance is never the same after classification; correlation is never the same after correspondence. The same goes for the forms of power that each enables. Classification may have enabled disciplinary power but sovereign power is never the same after classification enters the scene as a technology of power. As we seek to identify sensory power as a new form of power, we must also be aware that overt claims such as artificial intelligence, brain–computer interfaces and the construction of learning machines are also accompanied by subversive or infelicitous affects and effects.

These concepts are operative because each of them is characterised by operating or working, exerting force and active in producing the power to produce affects and effects, to produce objects and subjects. Concepts become operative by organising practices, instituting routines and arrangements and producing modes of behaviour appropriate to them. Data practices such as survey, census, inquiry and mining have different trajectories but there is also a thread that weaves them together. They are recognisable in each other even though they organise different methods. We have seen that the boundary between survey and census is that the latter embodies the former. Similarly, inquiry operationalises surveillance and classification and prepares the ground for the invention of correlation and correspondence. Today, the practice of mining, carried out by learning machines, embodies both concepts and practices.

The methods by which data practices transform numbers into knowledge have intertwined trajectories. The transformation from statistics to analytics between the 1890s and the 1930s coincides with the transformation from census to inquiry. The transformations of analytics into clustering coincides with transformations of inquiry into mining.

The form of knowledge we call datasciences includes all these concepts, practices and methods, as well as the ways in which objects and subjects are produced. If we approach the datasciences both through these parts that make them up (concepts, practices, methods, objects and subjects) and through the whole discourse that is generated not only by them but also through the forms of power that are embodied in each of their constituent parts, we can think critically about how social sciences and humanities can approach sensory power. It is absolutely necessary to explore the datasciences as a performative form of knowledge but it is not sufficient to understand sensory power. This is because sensory power encompasses another form of knowledge, neurosciences. The convergence of datasciences and neurosciences is the primary event that led to the invention of sensory power. In essence, it is only when doing things with numbers converges with doing things with neurons that we can observe the conditions of possibility for the invention of sensory power. We need to show how doing things with numbers converged with doing thing with neurons.

Notes

1. Alain Desrosières notes that 'the gap between technical objects and social objects – dating back to the seventeenth century – is now a deep one. From this standpoint statistical objects, whose good repute is linked to that of these two categories of objects, are admirably suited to a re-examination of this essential manner of dividing up the modern world; also, to a reflection on the political consequences of the fact that it is both

difficult and indispensable to conceive these objects as simultaneously constructed and real, conventional, and solid' (Desrosières [1993] 1998, 325).

2. 'The erosion of determinism and the taming of chance by statistics does not introduce a new liberty. The argument that indeterminism creates a place for free will is a hollow mockery. The bureaucracy of statistics imposes not just by creating administrative rulings but by determining classifications within which people must think of themselves and of their actions that are open to them. The hallmark of indeterminism is that cliché, information and control. The less the determinism, the more the possibilities for constraint' (Hacking 1981, 25).
3. When contemporary scientists aim to establish the boundaries of the new science called 'data science' resort to definitions that they tend to think as general (Tashman 2024, 1; Pietsch 2022, 5).
4. The history of computing from Charles Babbage to William Farr and beyond is provided by Haigh and Ceruzzi ([1998] 2021) and Russell and Norvig (2022). These histories often provide narrow accounts of the machine itself and how it is put together rather than social and political conditions of their possibility. Jon Agar (2003) is an exception because he provides a parallel and integrated account of the development of the computer as a machine and the government machine for governing people.
5. This scholarship is quite diverse but its origins can be traced to a research group of some twenty scholars organised by Lorenz Kruger in Bielefeld in 1982–3 (Hacking [1975] 2006, xii). In 2006, Hacking called the contributions of this small group an 'amazing little library' (Hacking [1975] 2006, xii). These include Porter ([1986] 2020). Hacking ([1975] 2006; 1990), Desrosières ([1993] 1998) and Stigler (1999).
6. We are using the phrase 'doing things with numbers' to draw inspiration from J. L. Austin's (1962) influential work on performativity but also to borrow from an interesting study of ancient Greek numeracy (Sing et al. 2022).
7. This scholarship has also inspired recent contributions that revise, amplify, shift and extend various aspects of their contributions. We also draw on this scholarship. The notable examples are Emigh, Riley and Ahmed (2016a; 2016b), Thorvaldsen (2018), Anderson (2015), Curtis (2001), Anderson (2015), Hald (2005), Schweber (2006) and Higgs (2004).
8. Longbing Cao (2018) provides one of the earliest comprehensive accounts of data science as such. The focus is on techniques and differentiating it from statistics and analytics by incorporating them into it.

William Cleveland (2014) provides a more general case for data science as a distinct field. Brian Godsey (2017) argues that thinking like a data scientist is different from thinking like a statistician or even data analyst. There are now numerous guides to data science (Tashman 2024; Zhu 2023).

9. Technical introductions such as those by De Brouwer (2020), Ogihara and Bonnell (2023) and Cremonini (2024) focus on specialised programming languages such as R and Python and numerous packages developed within them for analysis of large datasets. But these techniques also gradually define major epistemological transformations, as we will see below.
10. Foucault ([1997] 2003, 249–50) notes the emergence of disciplinary power in response to a problem of government: 'It is as though power, which used to have sovereignty as its modality or organizing schema, found itself unable to govern the economic and political body of a society that was undergoing both a demographic explosion and industrialisation. So much so that far too many things were escaping the old mechanism of the power of sovereignty, both at the top and at the bottom, both at the level of detail and at the mass level. A first adjustment was made to take care of the details. Discipline had meant adjusting power mechanisms to the individual body by using surveillance and training.'
11. As Stuart Elden (2017, 139) notes, 'the whole point is that discipline is made up of two elements – surveillance and punishment'.
12. In particular, scholars on security have explored the changing nature of surveillance (Lyon 2018; Bauman et al. 2014; Lyon 2014).
13. As Hacking (1981, 21) notes, 'Francis Galton began to explain biological facts, and the phenomena of human heredity, by deducing them from the normal or Gaussian distribution. Thus, what he called regression towards mediocrity (what we call regression towards the mean) was prompted by questions about traits of notable European families and led to our modern conception of the correlation coefficient. When we use a statistical law not only to predict and organise phenomena but also to explain them, chance is well on the way to being tamed.'
14. Alain Desrosières (2016) notes that in the early twenty-first century another form of psychometrics emerged and 'it provided the statistical tools for differential psychology, which was concerned with the statistical distributions of the properties of the human mind. Its founding document was a publication by Charles Spearman (1904), who extended Karl Pearson's ideas on correlation and discovered, within the multidimensional space of the abilities of one school's pupils, a general

factor in human intelligence, a predecessor of the intelligence quotient' (189). Desrosières says that in 1935 'Louis Leon Thurstone attacked the one-dimensionality of this notion and developed multidimensional factor analysis; the suggestion was that the human mind may be endowed with qualities that are incommensurable, thereby calling into question the one-dimensional scale that had dominated the eugenicist human sciences since Francis Galton (189–90).

15. Desrosières ([1993] 1998, 103–5) devotes an entire chapter to correlation and its transformation into something else without yet naming or identifying its logic. See Mackenzie (2017, 23) on the importance of this family of techniques in the development of machine learning.
16. Pearson ([1892] 1911, 124–5) says: 'When the exertion is at once determined by the immediate sense-impression (which we associate with a construct projected outside ourselves), we do not speak of will, but of reflex action, habit, instinct etc. In this case both sense-impression and exertion appear as stages in a routine of perceptions, and we do not speak of the exertion as a first cause, but as a direct effect of sense impression; both are secondary causes in a routine of perceptions. . . . None the less the inherited features of our brain, its present physical condition owing to past nurture, exercise, and general health, our past training and experience are all factors determining what sense-impressions will be stored, how they will be associated, and to what conceptions they will give rise. . . . Science tries to describe how will is influenced by desires and passions, and how these again flow from education, experience, inheritance, physique, disease, all of which are further associated with climate, class, race, or other great factors of evolution.' Desrosières ([1993] 1998, 109) quotes this passage and notes that it was written before 1892, when Pearson became involved in the eugenics movement.
17. Desrosières says that 'from this text—a surprising one, given the current image of a eugenicist and hereditarian Pearson (which he indeed became later on)—one can see how malleable are the reserves of theories and interpretations that a scholar can draw upon, and how dependent on a network of extended connections, in which elements habitually termed scientific, philosophical, social or political are all mingled: when he wrote this passage, Pearson did not yet know what path he would take. The one thing he was certain of was that he wanted to combat the idealism and metaphysical thinking of the old English style of university in favour of a scientific positivism embodied by the physicist Mach.'
18. Agar's (2003) discussion of how George Richardson Porter in binding the official and non-official wings of the expert movement of statisticians

together is illuminating. Agar documents the increasing role of government in the production of statistical facts and the controversies surrounding centralisation of statistical knowledge. The development of punched-card office machinery by Herman Hollerith and its application in government departments appears less an invention than a necessity. The use of punched-card machinery in the 1911 census and its connection to the expanding boundaries of the state. The punched-card machinery was also used in the National Insurance Act and the role of the Royal Statistical Society in providing statistics for government use. Agar's account of the fall from power of the expert movement of mechanisers and the impact of the introduction of large mainframe computers in government departments in the latter part of the twentieth century is the institutional background of the shift from statistics to analytics and the invention of datasciences. Agar concludes that the relationship between discursive and material mechanisation and the role of government as producers of knowledge is a key to understanding the relationship between technologies and power.

19. Foucault (2016, 233) was interested in the administrative survey in the seventeenth to eighteenth centuries not only because of their function but power was exercised through a number of people such as intendants, police apparatus, and so on and how this power gave rise to a number of forms of knowledge.' This was Foucault's persistent method where he moved from practices to methods, from concepts to practices, and from methods to concepts in a recursive way and always traversing their relations with different forms of power.
20. These friendly societies drew Foucault's attention and Hacking (1981, 23; 1983, 282, 290) provides further analyses of their function in the invention of regulative power. More recent studies in friendly societies have highlighted how the population emerged as an object of government (Cordery 2003; Leeuwen 2016). Agar (2003, 156) shows how the government machine incorporated the friendly societies.
21. The most notable specialist studies are Glass (1973), Kertzer (2002), Higgs (2004), Levitan (2011) and Anderson (2015). The most notable studies of the census in the development of population statistic are Stigler (1999), Hacking (1990), Porter ([1986] 2020) and Desrosières ([1993] 1998).
22. Wells (2015); Glass (1973).
23. Hacking (1990, 17) is quite certain about this. He says: '[I]n the modern era, however, a census was an affair more of colonies than of homelands. The Spanish had a census of Peru in 1548, and of their North American possessions in 1576. Virginia had censuses in 1642–5 and a decade later. Regular repeated modern censuses were perhaps first held in Atadie and

Canada (now the provinces of Nova Scotia and Quebec) in the 1660s. Colbert, the French minister of finance, had instructed all his regions to do this, but only New France came through systematically and on time. Ireland was completely surveyed for land, buildings, people and cattle under the directorship of William Petty, in order to facilitate the rape of that nation by the English in 1679. The sugar islands of the Caribbean reported populations and exports to their French, Spanish or English overlords. New York made a census in 1698, Connecticut in 1756, Massachusetts in 1764. The United States wrote the demand for a decennial census into the first article of their Constitution, thus continuing colonial practice, and even extending it, as westward the course of empire took its way, across the continent and in due course to the Philippines. Going east, the British took the same pains to count their subject peoples. India evolved one of the great statistical bureaucracies and later became a major centre for theoretical as well as practical statistics' (Hacking 1990, 17).

24. Thorvaldsen says that there was no direct link between the French census-taking in Quebec and the Danish census-taking in Norway in the 1660s. But 'they signalled a common interest in population control among mercantilists. These censuses must be viewed as colonial control mechanisms rather than embryonic attempts at nation building on either side of the North Atlantic. Denmark had recently brought Norway into a tighter union in connection with the introduction of enlightened despotism in the Danish realm from 1660. For Quebec it has been asserted that France had intended for its colony along the St Lawrence to be self-sufficient with respect to population growth since Colbert stated on behalf of French mercantilists that they did not intend to export people who were needed at home' (Thorvaldsen 2018, 19).
25. Evelyn Ruppert (2012, 209–10) says that, 'while the population census has been a key method and continues to be so, it is considered to have a number of shortcomings: it is too expensive, cannot deliver data sufficiently fast and at sufficient levels of locality and detail, and cannot be integrated with other data sources. Consequently, population registers and joined-up government administrative databases are being investigated or implemented as alternatives. Do all of these methods constitute surveillance? Yes and no. The reason is that rather than simply being different perspectives on a population that is "out there" each method identifies and makes up different populations. This is because they constitute their population objects (what) and subjects (who) differently.'
26. 'Of course, in many scientific practices measure, inquiry, and examination are put to work simultaneously as pure and simple methods or

strictly controlled instruments. It is also true that at this level and in this role, they are detached from their relationship with forms of power. But before appearing together and thus clarified within definite epistemological domains, they were bound up with the establishment of a political power; they were both its effect and instrument, measure fulfilling a function of order, the inquiry a function of centralization, and examination a function of selection and exclusion' (Foucault 2019b, 221).

27. As early as the end of the twentieth century, Jain, Murty and Flynn (1999) were signalling how cluster analysis could be transformed to meet the demands of data mining.
28. Jain, Murty and Flynn (1999, 310) indicate that 'data mining, like clustering, is an exploratory activity, so clustering methods are well suited for data mining. Clustering is often an important initial step of several in the data mining process. Some of the data mining approaches which use clustering are database segmentation, predictive modelling, and visualization of large databases.'
29. As Mackenzie (2017, 9–10) suggests, 'if we understand machine learning as a data practice that reconfigures local centres of power and knowledge by redrawing human–machine relations, then differences associated with machine learners in the production of knowledge should be a focus of attention' (Mackenzie 2017, 10).
30. 'Big data' is one name for this development. Yet putting the emphasis on 'data' rather than data practices and the need for understanding these data practices interviewing with concepts and practices that have stimulated them, and for understanding the form of knowledge and power that generated them, makes it a theoretically inadequate concept.
31. Pietsch (2021; 2022).
32. Kitchin (2014; 2021).
33. As Johns (2017) and Johns and Compton (2022) show, the infrastructures of data gathering, storage and retrieval constitute complex assemblages of law and territory. See also Sullivan (2022).
34. As Agar (2003, 79) argues, 'as statistics developed they also changed. In particular there was a profound shift from enumeration (How many poor are there?) to analysis (What regularities, laws even, are revealed in the statistics of the poor?). This shift was highly significant for the history of science, since it was not confined to the social sciences but shaped the physical sciences too, where statistical theories of gases, thermodynamics, and, later, matter (quantum mechanics) appeared.'
35. Desrosières (1991, 239) observes that 'This history of the idea of representativeness has brought us back to the question of generalisation: what

is the part, what is the whole. We have seen that the two operations, the political and the cognitive, in order to delimit a relevant whole (the nation or the market, whatever may be defined as appropriate) and to define and make comparisons possible between elements in this whole (citizens or consumers, as appropriate) were indispensable in making possible the appearance of the modem idea and technical implementation of representativeness.'

36. Hacking (1981, 25) says that statistics 'is all part of a transnational industrial philosophy that marks the very beginning of statistics. One candidate for the first "oeuvre" of statistics was The Statistical Survey of Scotland, a twenty-one-volume collection, 1791–1799. The ministers of the church of Scotland respond to detailed questionnaires about the state of their parishes. By the word "statistical" Sinclair tells us that he means "an inquiry directed at the conditions of life of a country, in order to establish the quantum of happiness of the inhabitants." Neither Sinclair nor the Calvinist ministers of the church of Scotland were Benthamites, but they were all part of a game that would establish a "calculus of felicity."'
37. Anders Hald (2005) argues that the history of probability and statistics and their applications before the eighteenth century shows that statistical methods and the state had overlapping trajectories. Much has been made of the fact that the origin of the Italian word 'statistics' is derived from the word 'stato', meaning state. The word was used in Italy in the sixteenth century and spread to France, Holland and Germany in the seventeenth and eighteenth centuries. It was taught in universities and consisted mainly of descriptions of the important characteristics of states, such as population, economy and geography. However, the systematic collection of numerical data on the population and economy of the Italian city states during the Renaissance shows that such data practices were already widespread before the emergence of the modern state. Hald shows how John Graunt's analysis of weekly reports of vital statistics for London in 1662 led to the development of analytical methods called political arithmetic, popularised by William Petty. The work of Caspar Neumann, who used registers of births and deaths to combat popular superstitions about health and sent his observations to the Royal Society in London. This led to Edmond Halley's analysis of the data and the publication of his paper 'An Estimate of the Degrees of the Mortality of Mankind' in 1693. The historical development of statistics, including the origin of the word 'statistics', the collection and analysis of numerical data and the early application of statistical methods to vital statistics and mortality rates, should not be reduced to state data practices.

38. As Porter ([1986] 2020, 243) says, Galton's eugenics was a programme for improving humanity through selective breeding rather than social reform. That is why 'he was interested not in mean values, but in the tails of the error distribution, where exceptional individuals were gathered'.
39. Ruppert and Scheel (2019) discuss some aspects of this competition.
40. As Anil Jain (2010) shows, clustering has a long history. K-means, a popular algorithm, was proposed in 1955 and is still widely used, highlighting the challenges in designing general-purpose clustering algorithms. Jain also shows how emerging research such as semi-supervised, ensemble and large-scale clustering are key developments to machine learning.
41. Jain, Murty and Flynn (1999, 265) were among the early pioneers indicating that 'clustering is useful in several exploratory pattern-analysis, grouping, decision-making, and machine-learning situations, including data mining, document retrieval, image segmentation, and pattern classification. However, in many such problems, there is little prior information (e.g. statistical models) available about the data, and the decision-maker must make as few assumptions about the data as possible.'
42. Clustering is a process of grouping data items based on a measure of similarity or proximity. As Jain, Murty and Flynn (1999, 315) observe, 'Clustering is a subjective process; the same set of data items often needs to be partitioned differently for different applications. This subjectivity makes the process of clustering difficult. This is because a single algorithm or approach is not adequate to solve every clustering problem. A possible solution lies in reflecting this subjectivity in the form of knowledge. This knowledge is used either implicitly or explicitly in one or more phases of clustering. Knowledge-based clustering algorithms use domain knowledge explicitly.'
43. Jain et al. (2004) show numerous clustering algorithms, their taxonomies and evaluation studies. They argue that different clustering algorithms deliver similar results. However, analysis of the similarity and properties of clustering objective functions must be evaluated from the perspective of the objectives of research. They illustrate how conventional clustering algorithms and more recent algorithms partition and group data. They conclude that only a small number of clustering algorithms are sufficient to represent a large spectrum of clustering criteria.
44. Tryon and Bailey ([1939] 1970) was a pioneering study. It is instructive to see the differences between the first edition in 1930 and the second edition in 1970. The former is more theoretical and the latter contains numerous applications. See Murtagh (2016, 26) and Hennig and Meila (2016).

45. As Desrosières (2016, 186) says, '[correspondence analysis] is more a technique for producing descriptive analyses than one used to prove or refute hypotheses in the orthodox Popperian sense, which has sometimes caused it to be discredited, particularly among economists. On the other hand, it met with great success in French sociology, being used by Pierre Bourdieu among others, as well as in history and political science.' Bourdieu ([1979] 1984, 261) describes its usefulness as it 'makes it possible to isolate, through successive divisions, different coherent sets of preferences stemming from distinct and distinctive systems of dispositions, defined as much by their interrelationship as by the relationship between each of them and its social conditions of production.'
46. The debate over how Bourdieu used correspondence analysis is intriguing as it reveals conflicting claims (Le Roux and Rouanet 2010, 4; Lebaron and Robson 2009; Martin 2009).
47. Bourdieu defines habitus as a system of durable, transposable dispositions that generate and organise practices and representations without conscious aiming at ends or express mastery of the operations. As we have seen, correspondence analysis, used by Bourdieu in his work, allows social scientists to organise data, observe similarities and differences between categories of variables, and graphically depict relationships. It is an exploratory method capable of sorting a multiplicity of variables into plots. It is similar to factor analysis and cluster analysis but focuses on determining complexity and identifying complex patterns of behaviour in relation to sociodemographic variables. Cockerham and Hinot (2009) argue, however, that correspondence analysis can only be used for displaying relationships, not hypothesis testing. They use other statistical techniques such as regression analysis and path or structural equation modelling for testing hypotheses and measuring relationships with predictive power exclusive of the effects of other variables. They claim the importance of class circumstances as the key variable for understanding the dispositions produced by habitus in Bourdieu's conceptualisation. Conventional measures of social class like income, education and occupation need to be viewed in combination with each other to constitute a structural variable that produces top-down distinctions in the quality and form of dispositions generated by habitus.
48. Luc Boltanski not only pays homage to Alain Desrosières for his contribution to the social sciences, particularly in the context of France in the years after 1968, but also articulates two key questions that were central to the social sciences in this period. The first question concerns the possibility of social science and how to distinguish the images of social

reality produced by the social sciences from those produced by social actors. Desrosières's work focuses on statistical reasoning as a technique and instrument of power, especially state power, and its role in shaping social reality. The second question concerns the relationship between the social sciences and social critique. Boltanski says that Desrosières's work explores how social sciences can be critical and serve as a science of the social, while at the same time engaging in social critique. The concepts of science and the state imply each other, but there is a need for a particular concept of science and a particular way of doing science, as well as a particular concept of the state and a particular way of relating to the state. Boltanski discusses the critique of statistical reasoning and the role of statisticians in underpinning a critique of the existing social order. For Boltanski, Desrosières fights to defend this position against those who used quantitative methods in the service of the state, arguing for the need for 'reflexive numbers' in modern societies.

3

Doing things with neurons: neurosciences

How did the brain become a subject of knowledge and an object of power? How did diverse disciplines and fields converged to create a form of knowledge called neurosciences? How did datasciences and neurosciences instituted an imaginary of the brain as a learning machine? The chapter is strictly genealogical: it cannot aim to provide even an outline of the development of neurosciences but aims to highlight how datasciences and neurosciences instituted a particular imaginary of the brain, set out to create a machine to fit that imaginary, and shaped a brain–machine regime. How has this imaginary generated strategies and technologies of power? We have seen above how clusters have become subjects and objects of power and knowledge with the development of datasciences. How did clusters emerge as transversal assemblages with constantly monitored, evaluated and transformed pulses functioning like an imaginary brain over the last 100 years?

This chapter first focuses on the invention of the brain as an object of knowledge and power through the intertwining trajectories between datasciences and neurosciences and then outlines how datasciences have converged with neurosciences to increasingly constitute clusters as learning machines becoming subjects and objects of government.[1] The first stage of the analysis involves the invention of the brain under the influence of datasciences. The second stage is the attribution of the functions of the imaginary brain to machines. The consequence of this is the transformation of clusters in datasciences into clusters in neurosciences with machinic qualities. It is this two-stage invention that produces a new imaginary of polities as learning machines, which we will further explore in the next chapter.

The last chapter examines datasciences with the concept of 'doing things with numbers'. We have shown that datasciences assemble discursive and non-discursive fields, including 'data science', but it traverses particular concepts, practices and methods and produces specific objects and subjects of government. It is the performative affects and effects of these assemblages that we refer to as doing things with numbers. This chapter examines neurosciences with the concept of 'doing things with neurons'. Similarly, neurosciences is also an assemblage of discursive and non-discursive fields, including 'neuroscience', but it traverses specific concepts, practices and methods that produce objects and subjects of government. Again, it is the performative affects and effects of this assemblage that constitute doing things with neurons.

The imitation game

The brain has a long past but a short history. We learn the former from histories of the brain that trace it back millions of years. The emerging field of neurohistory, for example, examines the role of the brain and its evolutionary development in shaping human history and culture.[2] A claim of neurohistory is that the human brain evolved over 1.7 million years to navigate the complexities of social living, and that many of our behaviours are shaped by dispositions, moods, emotions and feelings with deep historical lineages.[3] As Daniel Smail (2008) shows, neurohistory navigates this past through a rejection of genetic determinism or an essential human nature. As a kind of historical ontology, neurohistory makes effective use of the concept of the plasticity of the brain, where universal emotions like fear, pleasure and pain are linked to particular cultural events through malleable neural pathways. The ability to modulate emotions means that body states and automated responses do not rigidly dictate behaviour but rather provide a backdrop that people evaluate and make decisions against. Yet these dispositions and feelings still have a structural influence on human actions and history, even if they are sometimes overridden.[4] As Catherine Malabou puts it, 'the bond between brain and history – concepts long taken to be antithetical – is now established with certainty' ([2004] 2008, 1). Is the long past of the brain being interpreted by its short history?

The short history of the brain comes from the rapid advances in neuroscience technologies and theories in the late twentieth century generating a new imaginary of the brain's plasticity, modularity and connectivity in the twenty-first century.[5] If this new imaginary distances neurohistory from the deterministic ideas of innate, hereditary differences that had previously permeated the natural and human sciences, it raises the question how this short history has affected its long past. If neurohistory reveals that the biological life of brain involves its social life, shaped by the brain's plasticity, modularity, connectivity, in other words its historicity, which involves both culture and biology, and evolves in unpredictable ways, the new image of the brain becomes as malleable as the body.

Is it a coincidence that this imaginary of the brain's plasticity, modularity and connectivity (its historicity) emerged as the brain became the object of both intervention and imitation in the twentieth and twenty-first centuries?[6] The development of machines that imitate and simulate the imaginary of the brain since the 1930s has been remarkable. The brain–machine analogy has become so pervasive, and the convergence between datasciences and neurosciences so profound, that the imaginary of the brain as a plastic, modular and interconnected machine has driven not only a particular imaginary of the 'natural' brain but also built an 'artificial' one. Yet we are interested in developing the concept doing things with neurons rather than assessing the claims of this analogy.[7] That, at least, is the claim of this chapter. We argue that the new imaginary of the brain would be impossible without the entanglement of datasciences and neurosciences as forms of knowledge. This has implications for understanding forms of politics that we call datapolitics and neuropolitics, which are both affects and effects of sensory power.

The short history of the brain as a subject and object of neurosciences spans the early twentieth to the early twenty-first centuries. Just as in datasciences, we use 'neurosciences' as a plural term to indicate that there are fields with clinical applications and experimental or applied practices. Like doing things with numbers in datasciences, we are interested in doing things with neurons not as a metaphor but as performatives that simultaneously develop an imaginary of the brain and construct machines that imitate that imaginary. We will, therefore, say very little about doing things with neurons in clinical, experimental and applied neurosciences. While clinical neurosciences may perform brain research for purposes of diagnostic and treatment of illnesses arising from the central nervous system, experimental and applied neurosciences often push brain research in the direction of the imitation game: the ultimate aim of creating machines that imitate and even surpass the human brain imagined in a particular way. Thus, neurosciences include clinical neuroscience as practitioners themselves often cross these boundaries in their practices and engagements. We will, however, especially highlight computational neuroscience from our perspective of the birth of sensory power.[8] It is this field of knowledge that entangles the histories of the brain and the machine in the last 100 years or so and intensifies the convergence and integration of datasciences and neurosciences. Although the studies of the brain are traced to the nineteenth century in clinical neuroscience, it was during and after the Second World War that the entanglement of the brain and the machine began with experimental and applied neurosciences.[9]

This entanglement was expressed most famously by Alan Turing (1950) as 'the imitation game'.[10] However, what he meant by imitation has been interpreted rather widely, partly because of the test devised by Turing, which assesses whether a human observing a dialogue between a machine and another human can tell which is which. If not, the test concludes that the machine is as intelligent as the human (Warwick and Shah 2016). This is ingenious (if not playful) because it actually limits what is meant by intelligence to how it is performed in a given time, space and even domain of life. This was the idea of intelligence that Turing held more or less consistently throughout the 1940s and early 1950s.[11] We will focus below on the 'imitation game' as a concept that captures the brain–machine convergence analogy of the last 100 years, which will be the background against which polities as learning machines and the autopoietic subject would emerge.

When Turing (1936) writes on computable numbers, he is dealing with a mathematical problem. Yet this is also a problem for computation, especially digital computation, as it leads Turing to propose a binary system. Turing introduces the concept of computable numbers, which are real numbers whose decimal representations can be calculated by finite means. Turing defines computable functions, computable predicates and other computable entities, noting that not all definable numbers are computable. Turing is aware that Alonzo Church (1936) had already introduced this idea as 'effective calculability', though Church had defined it differently. Still, Church had reached similar conclusions about and the proof of the equivalence between 'computability' and 'effective calculability'. Turing gives a more explicit definition of computable numbers, stating that they are those whose decimals are computable by finite means. He argues that the operations used in computing a number include all those used in computing a number with a machine. Significantly, he introduces the concept of 'automatic machines', where the motion of the machine at each stage is completely determined by its configuration. Turing defines 'computing machines' that print two kinds of symbols, the first kind consisting entirely of 0 and 1. Turing then imagines a 'universal computing machine' that can be used to compute any computable sequence, not just programmed for a particular problem. At this point, neither Church nor Turing mentions the intelligence of such machines, or whether we can imagine them as thinking machines.

Yet, within about fifteen years Turing (1950) will ask the question 'can machines think?' The computability problem now expands to the imitation game. Turing thinks that 'digital computers' can be designed to mimic the

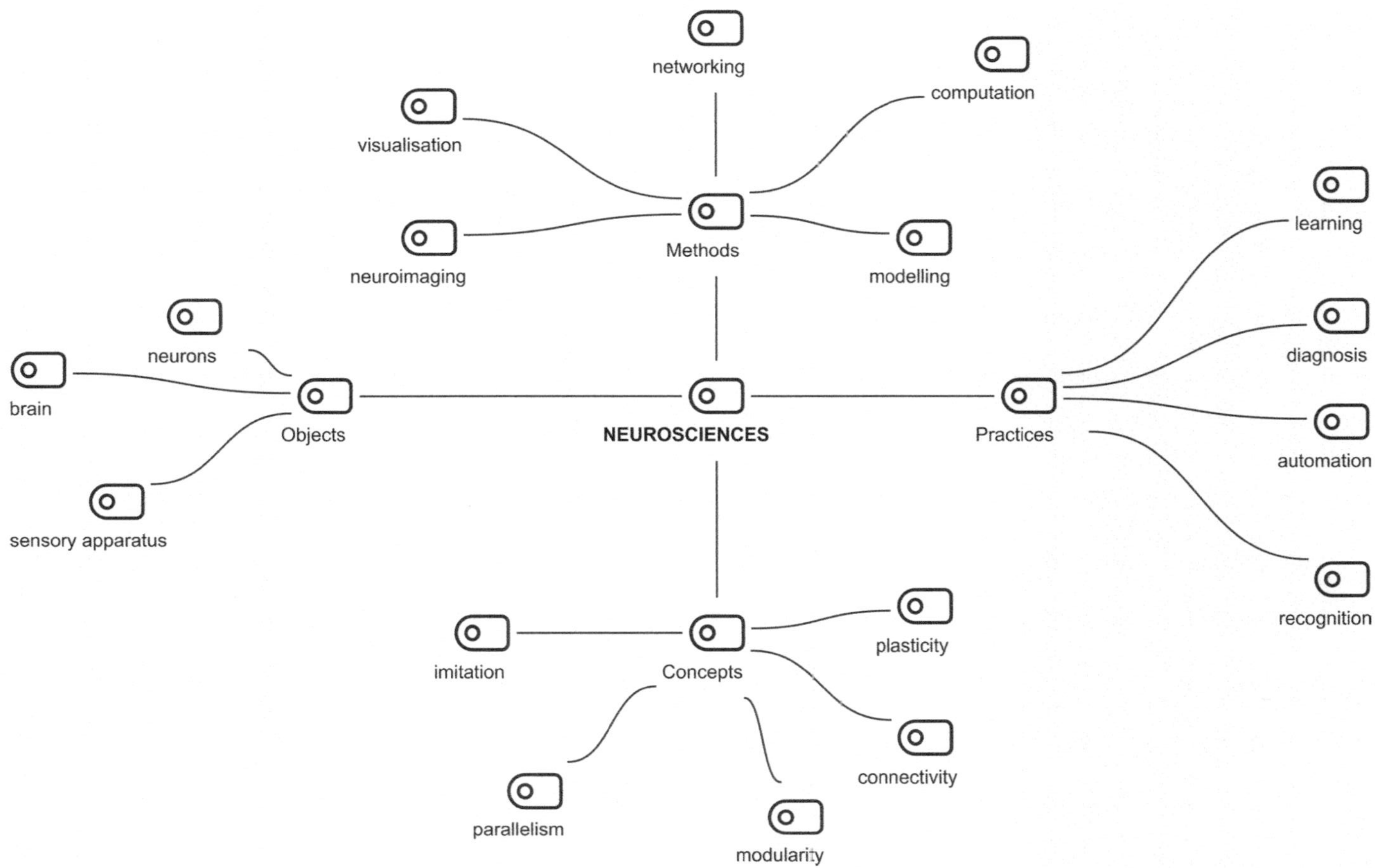

Figure 3.1 Neurosciences or doing things with neurons

behaviour of a 'human computer' very closely, and that the key question is whether there are imaginable digital computers that could perform well in the imitation game. Turing dismisses several objections to the idea that machines can think, including theological, mathematical and consciousness-based arguments. But his dismissal is as much about the idea as it is about the question. Turing suggests that the claim that machines can think is probably too ambitious, as he is not convinced that we can agree on what thinking is as an activity, or what intelligence involves. He suggests instead that it might be more pragmatic to try to create a 'learning machine', starting with a simple 'child-machine' programme and then subjecting it to a process of education similar to that which a human child develops. This would involve providing the machine with a store of information and logical rules, as well as a means of learning through rewards and punishments. Turing acknowledges that there are significant challenges in creating such a machine, but he expresses optimism that it may be possible. The learning machine is not intelligent but it is sentient (perceptive and responsive); it is capable of learning and acting on sensory stimuli it receives, and it perhaps has initiative. Turing elaborates on this idea in further articles and lectures ([1947] 2004; [1951] 2004).

What happened between 1936 and 1950?[12] The histories of early computers, the military involvement in their development, the role of influential figures such as Warren McCulloch and Walter Pitts (1943), Claude Shannon (1948), John von Neumann ([1958] 2012), Norbert Wiener ([1948] 1961), and Arturo Rosenblueth (1950) have been written about this decisive period. Almost invariably most narratives consider this period as the birth of the computer with various models in operation by the 1950s. Nevertheless, from the point of view of the birth of sensory power the genuine significance of this period is the entanglement between datasciences and neurosciences for developing the machine built on an imaginary of the brain. It is at this point that it becomes impossible to distinguish whether the brain constitutes an imaginary of the machine or the machine constitutes an imaginary of the brain. The key figures of this period involved in the development of the machine are deeply involved in and knowledgeable about the latest developments in neuroscience, especially neurobiology and neuropsychology. The subsequent naming of this period as the emergence of artificial intelligence is decisive, as there were deep differences among these figures, who were academics, clinicians, experimenters and practitioners. They were divided over whether it was possible, or even desirable, to attempt to create machines as brains. The entanglement of brain and machine was more than an analogy. Most believed that it was possible to create learning machines that imitated the characteristics of nervous systems but to imagine them as analogous to the brain was not a commonly shared assumption.[13]

All of this has developed in a rather staggering way between the 1940s and the 2020s, rapidly at first (1940s–1960s) and then more slowly (1960s–1980s), then immensely (1980s–2000s), then exponentially (2000s–2020s), artificial intelligence producing learning machines for the apparatuses of war, finance, education, health, cities and climate (Russell and Norvig 2022). For all this to happen, a number of things have to come together. It is beyond the scope of this book to discuss that history further. As in Chapter 2, we will focus on the key concepts, practices and methods that characterise these developments, and the key moments of convergence between the datasciences and the neurosciences. George Dyson (2012, ix) captures our task well when he says that 'the stored-program computer, as conceived by Alan Turing and delivered by John von Neumann, broke the distinction between numbers that mean things and numbers that do things. Our universe would never be the same'. Yet this leaves out doing things with neurons, which may have been the most decisive transformation.[14]

Concepts: imitation, parallelism, modularity, connectivity, plasticity

The number of concepts at play in neurosciences is, of course, immense. It is possible, however, to select a few that dominate the overall episteme of neurosciences and its influence beyond medicine towards social and political arrangements. We will begin with imitation, which encapsulates how neurosciences have been entangled with the brain–machine analogy since the 1930s. It is this entanglement that has produced the concepts of parallelism, modularity, connectivity and practices that now permeate both neurosciences and datasciences.

Imitation

We must return to the origins of the imitation game before it was named by Turing. As we have mentioned, it was between 1936 and 1950 that the analogy between the brain and the machine as objects of knowledge and power was established. It was also in this period that the convergence between datasciences and neuroscience began, and the brain and the machine increasingly moved from being an analogy to combined and integrated developments. That was when their trajectories became entangled. Turing (1936) and Church (1936) had already made binary as the foundation of digital computing. Arguably, the breakthrough came when McCulloch and Pitts (1943) connected the binary system and the nervous system. To put it differently, when McCulloch and Pitts argued that neurons transmitted information through a binary system, and that the brain and the machine were analogous, they effectively transformed the imitation game into a research and applied game, converging datasciences and neurosciences.

McCulloch and Pitts present a logical calculus for understanding the activity of the nervous system, based on the premise that nervous activity exhibits an 'all-or-none' character, which allows it to be treated using propositional logic. They outline several key assumptions about the structure and function of the nervous system, including the presence of neurons with soma and axons, synaptic connections, thresholds for excitation and the propagation of impulses. They also include the concepts of inhibition, facilitation, excitation and learning and how these can be incorporated into the logical system. They then describe two types of neural networks: those with and without circular paths. For the former, they acknowledge the increased complexity due to the possibility of reverberating activity in the circular paths, which can lead to indefinite references to past events. For the latter, they provide a detailed demonstration, including the physical assumptions and the logical implications. They illustrate their awareness of the broader implications of their theory, particularly in relation to causality, knowledge and psychology. They argue that their theory highlights the inherent incompleteness and indefiniteness of our knowledge of the world, including ourselves, due to the abstraction and simplification required to make our knowledge useful. They conclude that the 'all-or-none' nature of nervous activity and its conformity to propositional logic have important implications for psychology, as all mental events can be seen as having an intentional or 'semiotic' character. They see that the production of learning machines can have implications for psychiatry by providing insights into the underlying neural mechanisms since the inference from overt behaviour to specific neural networks is not unique, and the unpredictable nature of neural activity can pose challenges for prognosis and treatment.

This article became the foundation for conferences held between 1946 and 1953 that developed analogies between nervous systems and computer systems. After the war Warren McCulloch, under the sponsorship of the Macy Foundation, initiated an interdisciplinary meeting of about twenty interested scientists to help launch explorations of the emerging analogy between the brain and the machine (Heims 1980, 202). These conferences were preceded by a broader conference series called Macy conferences, but the narrower series was named 'cybernetic conferences' to concentrate on the imitation game. The cybernetic conferences included neurophysiologists, neurobiologists, mathematicians, anthropologists and psychologists. Key figures such as Warren McCulloch, Claude Shannon, John von Neumann, Norbert Wiener and Arturo Rosenblueth were regularly present, although von Neumann and Wiener emerged as two central figures representing contrasting uses of the imitation game (Heims 1980; Conway and Siegelman 2005).

John von Neumann as a prominent mathematician became a fervent supporter of the military-industrial complex and its technological research and development. He went as far as advocating a 'preventative' atomic attack against the Soviet Union during the Cold War and worked closely with the military on weapons development, including the hydrogen bomb.[15] By contrast, Norbert Wiener, another prominent mathematician, refused to do research that could be used by 'irresponsible militarists'. The two mathematicians also had contrasting views on the feasibility of creating formal-logical models of the brain. Von Neumann was fascinated by the organisation of the human brain and envisioned a mathematical or logical system to describe its biology, although he acknowledged that the complexity of the nervous system posed significant challenges. Wiener, on the other hand, questioned von Neumann's assumptions about human decision-making and behaviour, arguing that people act for rational and irrational reasons, often based on emotion and misinformation. Nevertheless, Wiener and von Neumann were both deeply involved in the emerging field of cybernetics, attending the same conferences and discussing the topic together extensively. Their relationship became increasingly strained, with Wiener openly expressing his disapproval of von Neumann's military work. Von Neumann's contributions to computer science were also significant. Both were influenced by Alan Turing's work and, in turn, influenced Turing's work. Von Neumann created the von Neumann machine, which became the fundamental architecture for modern computation, which led him to explore the similarities and differences between the brain and the machine, anticipating the accelerating pace of neurotechnologies and their profound implications for social and political organisation. Although these two mathematicians represent two different trajectories of the imitation game, they share its basic principle that the machine can be modelled after the brain and the brain has machinic qualities.

John von Neumann articulates his view in *The Computer and the Brain* ([1958] 2012), which advocates the production of computers as learning machines capable of mimicking and reproducing human intelligence – a vision that became known as artificial intelligence. Norbert Wiener expresses concerns about artificial intelligence vision for its militarism and corporatism and becomes an advocate for cybernetics used for health and education research, which he outlines in *Cybernetics* ([1948] 1961).

Von Neumann plays the imitation game by comparing computing machines and the human nervous system, highlighting their similarities and differences. He describes the fundamental principles of digital and analogue computing machines, as well as what makes machines digital and their ability to achieve high precision. Von Neumann shows considerable knowledge of neurobiology and neurophysiology to examine the digital characteristics of the nervous system, arguing that the basic component, the neuron, functions in a digital manner by generating and propagating nerve impulses. He compares the size, speed and energy consumption of natural and artificial components, concluding that natural componentry favours highly parallel automata with more, but slower, organs, while artificial componentry tends to favour fewer, but faster, serial organs. Von Neumann acknowledges the complexities of the nervous system, considering the role of stimulation criteria and the potential for analogue or mixed functioning within the predominantly digital system. He also shows his awareness of the latest research on memory within the nervous system, estimating the probable capacity required and suggesting various physical embodiments. He then proposes the concept of codes, analogous in both the brain and the machine, distinguishing between complete codes that enable a machine to solve specific problems and short codes that allow one machine to imitate another. The importance of both logical and arithmetical parts in the nervous system and the challenges posed by the low precision levels observed in the system's statistical message-transmission mechanisms become his main concerns. He concludes that the logical and mathematical structures employed by the nervous system are likely to differ considerably from the forms we are accustomed to in our conscious experience. Yet this does not discourage his conviction that a brain–machine imitation is possible.

Wiener's work on cybernetics and its links to the functioning of the human brain and nervous system is not fundamentally different from von Neumann's. He also explores the parallels between computing machines and the brain, particularly in terms of memory, learning and logical processing. Wiener accepts that the brain, like a computing machine, operates on an 'all-or-none' principle, with neurons either firing or resting. He proposes that information can be stored in the brain by changing the permeability of synapses, much as data can be stored in computers by changing the properties of certain elements. Wiener became interested in the concept of conditioned reflexes and how they could be implemented in computing machines to enable a form of learning. Although Wiener is concerned about the potential dangers of advanced computing machines, he is convinced that the energy efficiency of the brain compared to other computing machines, as well as the potential for self-replicating and self-learning machines, will lead to autonomous, if not intelligent, machines. The deep connections between the workings of the human brain and the principles of cybernetics and computing are presented as a version of the imitation game.[16] The question that we want to address is not whether machines and brains are analogical and whether they can be made to think alike but what these claims make possible especially in both imagining human and non-human groups (clusters) as self-organising assemblages (polities) designed to achieve certain objectives.

These questions were addressed by again by two contrasting visions of learning machines for governing polities. While Karl Deutsch sees cybernetics as a method for governing polities as self-steering or self-controlling machines in *The Nerves of Government* ([1963] 1966), Norbert Wiener presents a dystopic vision in his *The Human Use of Human Beings* (Wiener [1950] 1989).

Deutsch sees the nerves of government as the application of communication and control theory, particularly cybernetics, to the analysis of how power functions. Deutsch notes the growing interest in quantitative data and computer simulation in political research, as well as the merging of communication and control theory with general systems theory. He suggests that the development of communication and control theory is driven by the rise of electronic communications technology, the need to reformulate research problems in more empirical and mathematical ways, and a sense of transition to a new era of human history. For Deutsch, cybernetic or self-controlling machines such as modern radar tracking and computing machines, which can 'sense', 'interpret', 'predict', and 'act' on their physical or social space, parallel the functioning of the brain. He weaves the concepts of memory, recognition, learning and purpose in such self-controlling systems, and proposes an operational approach to understanding the notions of quantity and quality. Deutsch makes use of the concepts of feedback, equilibrium and learning, highlighting how feedback theory offers improvements over conventional equilibrium analysis. Deutsch outlines the implications of these concepts for understanding political power, social transactions and the capacity for self-transformation in polities. He suggests that the development of communication and control theory may have normative implications and raises questions about the value assumptions underlying this approach. Yet Deutsch certainly remains convinced that the concepts of autonomy, integrity and value in self-steering or self-controlling networks will retain their meaning and function. Just as he imagines the brain as a self-sustaining and self-controlling physical process that includes the operations of selecting, abstracting, communicating, storing, subdividing, recalling, recombining, critically recognising and reapplying information, he also sees polities the same way. If the parallels between individual and social brains have implications for understanding group learning and creativity, the role of feedback, goal and purpose for exercising power will involve the concepts of lag, gain and lead as measures of a polity's performance. Deutsch then explores the dimensions of political growth, including the growth of people, economic resources, operational reserves, autonomy and goal-changing ability, as well as the interplay between the growth of the polity and the growth of individuals and subgroups within it. Although assumptions about polities and power are pervasive in the imitation game, Deutsch's vision is among the most ambitious if not influential visions of how power is exercised in cybernetic polities as learning machines: generative, self-controlled, intelligent and yet programmable clusters with logical and sensory apparatuses.

Although written more than ten years earlier than *The Nerves of Government*, Wiener's *The Human Use of Human Beings* is a prescient warning about the cybernetic polities as learning machines. Let us make no mistake. It is not that Wiener does not believe in their possibility. He begins by discussing the shift in physics from the Newtonian view of determinism to the recognition of an element of chance and probability in the universe, as introduced by contemporary physicists. This recognition of incomplete determinism is paralleled in Sigmund Freud's (2001) admission of a deep irrational component in human conduct and thought. If, Wiener argues, cybernetics is the science of control and communication in both living organisms and machines, we must understand the difference between programmed behaviour and contingent behaviour. He emphasises that society can only be understood through a study of the messages and communication facilities that belong to it, and that the future development of communication between humans, machines and between machines will play an ever-increasing role. Wiener understands the differences between the prearranged behaviour of machines and the contingent behaviour of humans and animals analogous but different modes of behaviour. He knows that modern machines possess sensory apparatuses and can be conditioned by their relation to the external world, unlike the older, closed clockwork machines. The concept of feedback, where the performed action on the outer world is reported back to the central regulative apparatus, is essential to both living organisms and machines. Wiener draws the implications of this parallel between the physical functioning of living individuals and the operation of communication machines. He explores, for example, the concepts of rigidity and learning, where the mechanical fluidity of the human being provides for almost indefinite intellectual expansion, unlike the mechanical rigidity of the insect. For Wiener, specific learning machines and their potential future includes the possibility of artificial lungs and chess-playing machines. But he sees the sinister possibilities of such learning machines, particularly in the context of military and corporate applications. Eventually for Wiener the critic can bring inhibitory impulses to understand the nature of humans and their excitatory capabilities in order to wield this knowledge as soldiers, statesmen and entrepreneurs to control the use of machines. Wiener's conclusion that the dangers of using human beings as cogs and levers in vast organisations as learning machines raises the spectre of asking the right questions to receive the right answers.

Today's debates on artificial intelligence both mirror and embody the imaginaries produced by von Neumann, Deutsch and Wiener and are as assertive as those statements expressed by them. While von Neumann and Deutsch can be seen as seduced by the possibilities of the sensory power and Wiener as a critic of its uses, they were as confident about cybernetic movement as Pearson's seduction by the eugenic movement and the possibilities of regulative power. By overlooking the different forms of power and the differences between power as such and power as performatives, these claims produce the illusory that power is exercised for good or evil by the intentions or purposes of the 'holders' of power. This 'illusory' masks the question that the imitation game, any more or less than the population game, raises whether it can be played from the inside. The playbook of the imitation game could not have been further than the population game: while the latter was focused on calibrating bodies with parameters of a healthy and wealthy population (a relationship between the parts and the whole), the imitation game is focused on producing the autopoietic bodies responding to impulses forming dynamic learning machines (clusters) as polities. There are four concepts that especially encapsulate the development of the imitation game and its playbook since the 1930s and especially since the 1980s: parallelism, modularity, connectivity and plasticity.

Parallelism

The concept of parallelism has rendered the imitation game much more pervasive. Its emergence as a decisive concept in neuroscience and its implications for understanding intelligence and artificial intelligence has generated an imaginary of the brain. The brain is described as a highly parallel system, with many parallel streams of input and processing, in contrast to the sequential nature of the early digital computers or in fact the early images of the brain (Hassabis et al. 2017, 246–7). This parallelism is seen as a key feature of the brain's architecture and information processing, allowing for rapid decision-making and robust representation of concepts. It was in the 1980s that the development of parallel distributed processing and connectionist models motivated this understanding of the brain's parallel nature, and these models have had a significant impact on the field of artificial intelligence (Churchland and Sejnowski [1992] 2017, 59, 413).

Although parallelism has posed challenges as certain operations by machines can only be performed sequentially, and serialising a parallel procedure may impose new problems, which was recognised by von Neumann ([1958] 2012, 51), the drive towards understanding and incorporating parallelism in efforts to model and simulate intelligent machines, whether biological or artificial, gained a new momentum in the 1990s (Bielza and Larrañaga 2020, 30).[17] To put it differently, if the possibility of parallel and distributed computing was stimulated by the miniaturisation of computers in the 1980s, the possibility of modular and network computing can certainly be attributed to the development of the internet, or rather its extension from military, research and corporate domains into commercial and individual uses in the 1990s. The increasing networking of computers has made possible massively distributed parallel computing with modular and neural networks. It is not surprising that the emergence and development of these concepts were simultaneous in neuroscience and computer science, soon to be rebranded as computational neurosciences, which we will discuss later.[18] The importance of parallelism is in enabling two related concepts associated with intelligence, modularity and connectivity. These features influenced each other for the generation of new analogies between the brain and the machine and indeed blurring the differences between the two in the concept of singularity.[19]

Modularity

For neuroscience the brain increasingly exhibits a modular organisation, where it is decomposed into interconnected subsystems or modules.[20] These modules are defined at different scales, such as spatial regions of the neocortex, and short-term and long-term memory in the temporal regions. Within-module elements are more highly connected than between-module elements, providing compartmentalisation and reducing interdependence, which enhances the central nervous system's robustness. The hierarchical and modular structure allows for the formation of complex architectures with a high degree of functional specificity, enabling behavioural adaptation as each module can function and change without negatively affecting the rest of the system.[21] The modular architecture of the brain has increasingly been described using complex network theory applied to neuroimaging data, revealing the functional and structural hierarchical modularity of brain connectivity. Within these modular structures, brain regions perform different roles, acting as hubs with higher connectivity or as local processors, which are evident in both structural and functional connectivity networks and may have neurophysiological correlates.[22] The concept of modularity expresses that each region displays distinct patterns of energetic activity and maintains different trajectories of synaptic development, redevelopment and plasticity.

The concept of modularity also appears in the production of learning machines, where it can be expressed as a flat, modular or hierarchical structure. A hierarchical decomposition is important for reducing the complexity of building an intelligent machine that operates in a complex situation. Whether some neuroscientists accept the idea of a modular brain, encompassing components like the five senses, language and other cognitive functions, some see challenges in proving whether these modules are 'programmed' by genes rather than by culture. If indeed intelligence is seen as solely a result of neural activity, brain–body chemistry interacting with the nervous system becomes a problem. The concept of modularity illustrates how the languages of datasciences and neurosciences (especially computational neuroscience) become nearly interchangeable.

Connectivity

The modularity of the brain would only make sense if it were understood how neurons form specialised clusters to perform sensory and somatic functions of the body. If neurons are understood as forming networks that connect them with dendrites, axons and synapses, it becomes necessary to understand a brain, or a machine, as composed of neural networks, an interconnected system inspired by the arrangement of neurons in the nervous system; a program, a configuration of microprocessors, can be designed to imitate this. The structure and organisation of the brain, understood particularly as a network, provide new descriptions of cognitive processes and brain function. It then appears that the brain is not a simple collection of individual neurons, but rather a complex system of interconnected networks that exhibit emergent properties beyond those of single neurons.[23]

The converging points are that the brain is organised into hierarchical levels, from molecules and cells to networks, systems and behaviour, all of which contribute to cognition; that neurons form interconnected networks that enable information-processing capabilities beyond what a single neuron can do; that networks have specialised architectures and information-processing capabilities, which are integrated into larger brain structures and systems, enabling more complex cognitive functions; that techniques like brain imaging and functional connectivity analysis are crucial for understanding how different brain networks interact and contribute to various cognitive and emotional processes; that the social brain hypothesis suggests that the evolution of large brains in humans and primates is linked to the complexity of social skills, with specific brain regions like the frontal lobes, amygdala, and temporal lobe playing key roles in social processing; and that social emotions, such as embarrassment, shame and pride, involve the interaction between individuals and require the presence or imagined presence of others, highlighting the social nature of certain cognitive and emotional processes.[24]

If indeed the brain was imagined from the 1990s and 2000s as a neural network organising complex cognitive and social behaviours, what are the consequences of this for the emergence of the internet as a network of networks integrating brain and machine?[25] Malabou ([2004] 2008, 42) would say that 'there is no longer a centre but rather discrete assemblies of neurons forming mobile and momentary centres on each occasion. Organizational suppleness now goes hand in hand with authority and decision.' If we 'find ourselves faced with a complex organization that no longer proceeds in a top-down fashion from transmission to reception to retransmission of information but functions according to different, extremely complex, interpenetrating levels of regulation' then we will increasingly see cerebral organisation as connected networks of neurons (Malabou [2004] 2008, 43). Since there is no privileged vantage point, the connectivity of the brain and the machine will not be centralised and centralising. It is perhaps in this sense that cybernetics has always emphasised the self-controlling, self-steering and self-regulating machines. If the connectivity of learning machines is not merely their connectedness but their key architecture, then polities imagined as learning machines will have this architecture. As William Lytton (2002, 105) puts it, architecture of a network means its connectivity structure.

Plasticity

We have left the discussion of perhaps one of the most influential concepts to emerge from neurosciences in the twenty-first century to the end, not because it is the least or most important but because we think its importance in the birth of sensory power makes most sense against the background of the interrelated concepts of parallelism, modularity and connectivity that we have already discussed. If plasticity describes the brain's ability to change and adapt in response to experience and learning, it is another way of recognising its historicity.[26] This plasticity operates on three key levels: developmental plasticity (the formation of neuronal connections), modulational plasticity (the modification of neuronal connections through learning and memory) and reparative plasticity (the brain's capacity for regeneration and compensation after injury). Plasticity emerged as a fundamental property of the brain that enables it to be malleable, formable and self-cultivating. It contradicts rigid determinism, allowing the brain to evolve and transform itself through interactions within the social space. This plasticity is what enables humans to learn new skills, form memories and change their behaviour and cognition over time. Obviously, plasticity has important implications beyond neuroscience. It suggests a model of the brain and power that is not rigidly determined but rather open to transformation through recognition, domination and emancipation.

Could plasticity be used to justify economic, political and social systems focused solely on efficiency and adaptability rather than human flourishing? If, ultimately, the concept of plasticity reveals the brain's remarkable capacity for change, and how human genetic history and cultural embodiment intersect to shape people as unique yet social beings, can there be programmes for designing ontogenetic change and even more ambitiously can there be learning machines that incorporate plasticity as their operative principle? Malabou ([2004] 2008; 2022) rightly asks these questions when she emphatically calls neuroscience into question about what to do with our brains. Yet she assumes that power can directly intervene in our brains without considering practices through which learning machines are created and how we become integral components and elements of these machines.

Although these questions lie at the heart of sensory power they are often answered or modified in operational practices in the performance of this power. It is these practices that continually provide affective and effective answers to these questions, pushing the boundaries of the possible in governing diverse domains of life such as war, health, finance, education, climate and the economy.

Practices: diagnosis, recognition, automation

The concepts we have identified above – parallelism, modularity, connectivity and plasticity – are the entangled elements of different practices in playing the imitation game that enables both the construction of learning machines and the imaginary of the brain that drives them. These concepts do not merely mobilise practices of designing, building, producing learning machines; they are also continually repeated, iterated and resignified in doing things with neurons. Three practices have been influential in shaping and mobilising learning machines.

Diagnosis

Although advances in machine learning and deep learning have transformed medical diagnosis and most people become aware of learning machines through the advances in medical diagnosis, the concept of diagnosis is a broader practice of the learning machines. It is now common that sleep disorders are increasingly diagnosed with patients undergoing overnight observation in a sleep clinic, where their brain activity and muscle movements are monitored using electrodes (Churchland and Sejnowski [1992] 2017, 12). As learning networks trained on large datasets of patient records can achieve high accuracy in detecting certain medical conditions, diagnostic practices spread from health to other applications such as finance, economy and climate. Since combining the predictions of deep learning networks with human expert judgement results in even higher accuracies, the integration of two practices (deep learning and expert judgement) become broader. While deep learning and other artificial intelligence practices are advanced to improve the diagnosis of a wide range of medical conditions by leveraging large datasets and complementing human expertise, diagnostic practices replace, displace or complement conventional practices of disciplinary or regulative power such as description, explanation or prediction. Analysing data without assumptions and training learning machines without supervision increasingly affect judgement and value in diagnoses of the domain in which learning machines become prevalent outside medicine such as welfare, crime, policing, finance and migration. These diagnostic practices have not only epistemological (how we know) and ontological (what we sense) consequences but also political and ethical implications as they intensify the tension between description, explanation, prediction and diagnosis as regulative and sensory practices governing people and their conduct. The instantaneous and continuous evaluation and judgement that follows diagnostic practices generate decisions by intelligent agents on behalf of humans. We can say with a high degree of confidence that diagnostics is the decisive analytical component of sensory power as it emerges from the interface between datasciences and neurosciences: diagnostic practices create clusters as governing subjects and objects.

Recognition

The recognition of patterns in visual, numeric, aural, taste and smell data in neuroscience, artificial intelligence and computer science has evolved from shared origins in datasciences. The emergence of methods, especially cluster analysis and correspondence analysis, became influential in developing models of pattern recognition. That recognition became a key practice in computational neurosciences is associated with the drive to imitate the sensory apparatus of the human nervous system renders recognition as one of the most pervasive practices of regulative power. It is also the practice that makes diagnostics an operative practice of sensory power. The neural basis of cognition has been related to various brain projects to build intelligent machines to recognise objects and their physical and social spaces imitating human sensory apparatus. This may be considered the first step in the development of the brain–machine interface for artificial intelligence. Machine learning is essentially about building experience, with applications in pattern recognition and data mining: deep learning networks with millions of model neurons and billions of connections have succeeded in tasks like speech recognition and object recognition in images, surpassing human capabilities in some areas. Although this has contributed to medicine, prosthetics and surgery, most of its applications have been driven to triangulate memory, recognition and measurement to build learning machines capable of diagnostics and decision (Conway and Siegelman 2005, 343). It is often argued that information storage and memory are physical processes, but the recognition of quality involves establishing structural correspondence between the recognising system and the recognised system. The definitions of quantity and quality are introduced as essential components of recognition algorithms. These algorithms are essentially the same as supervised and unsupervised clustering algorithms we have discussed in the development of machine learning in datasciences. The meaning and function of recognition here means being able to identify patterns and their relations with each other within a specific physical space. As learning machines become more sophisticated (from repetition and signification) they also become increasingly adapt in recognising not only patterns and their relations but also their physical space. To put it differently, a learning machine evolves from a supervised agent into an unsupervised agent capable of drawing out inferences and diagnoses. Diagnostics are used to identify problems and not only suggest solutions by reasoning in certain domains but also perform actions. That travel agents, for example, are used to determine customer preferences and trade-offs to book travel arrangements on their behalf may appear a benign practice but we can also imagine how a citizenship practice such as voting can be delegated to learning machines (Poole and Mackworth [2010] 2023, 30).

As learning machines increasingly recognise not only objects and their relations but also the structure of social space, it has advanced claims to a new robotics development. While ethical and political concerns have been raised about the use of robotics in autonomous weapons, surveillance, policing and control, as these robots can be used to govern people and groups without human intervention, a development associated with sensory power may have much more pernicious and pervasive affects and effects: intelligent agents. The history and development of speech recognition, object recognition, language translation and sensory recognition signal that computational neurosciences are participating in the development of learning machines that combine diverse intelligent agents that embody both humans and non-humans.

Automation

As in plasticity, we have not left automation in practices in neurosciences to last not because it is the least or the most important but because the emergence of autonomous machines make most sense against the background of the proliferation of diagnostic and recognition practices. Automation and autonomous machines with significant implications across various domains are the most ubiquitous practices of learning machines. The proliferation of sensors is driving the trend towards automation, enabling continuous data collection and analysis for defined dimensions rather than specific individuals or moments. This has raised concerns about the potential for autonomous machines to mutate and replicate without human control. Norbert Wiener was approached by General Electric to advise on automation, but he refused, extending his boycott to any work that might hasten the deployment of automated machines and the displacement of workers (Conway and Siegelman 2005, 245). As we have seen, Wiener's concerns centred on the spectre of autonomous machines that could operate without human supervision or control and without ethical or political principles. The development of cybernetic models and the application of concepts like information theory, coding theory and self-steering systems have been slow, but development in these areas is accelerating the learning machines. The integrity of self-steering or self-controlling systems can be modulated by imposing non-autonomous changes or disrupting their functioning through various mechanisms. The growth of large military, corporate and state organisations is characterised by the interplay between the dimensions of growth of the organisation and the growth of the autonomous subgroups that compose it. The autonomous growth and self-determination, self-control or self-steering groups became elements of governing such organisations generating an entirely different model of organisation than the twentieth-century bureaucratic organisations. Autonomous machines are being deployed in various applications, including medical diagnosis, educational performance, factory scheduling, transportation logistics, routing, locomotion, autonomous planning and scheduling, machine translation, recommendations and even game playing. The development of legal and ethical standards for such critical applications has become intense issues in the development of learning machines. If learning machines become powerful intelligent agents for the struggles over cybersecurity where automation can be used to probe for vulnerabilities, conduct cyberattacks and eliminate anomalous behavioural patterns, then learning machines as polities developing such capabilities now becomes a critical challenge.

While these developments raise concerns about biased decision-making, impact on employment, safety-critical applications, privacy concerns and cybersecurity vulnerabilities, from our perspective of the birth of sensory power the most important thing is the logics of the agent-based approach to learning machines and the concept of autonomous computing, where computer systems, networks and learning machines monitor and control themselves, produce diagnostics and generate new goals and modes of conduct into their repertoires. The meaning of autonomy in autonomous systems must be sought in their performance rather than what is said about them by their advocates or critics. We would, in other words, need to understand methods by which the practices of diagnostics, recognition and automation are made possible. The emergence of methods of studying brain increasingly rely on datasciences and methods such as computation, modelling, visualisations and learning play a significant role in neurosciences, so much so that the emergence of computational neuroscience is considered a breakthrough. Discussing meaning of autonomy in autonomous machines will be possible only after we understand some of these methods.

Methods: computation, modelling, visualisation, learning

We begin with the methods of computational neuroscience to understand the outstanding affinity between the methods developed in data sciences in the last eighty years. A typical textbook on data science and a textbook on computational neuroscience almost cover identical methods, differing only in their objects.

Computation

If computational neuroscience has become an enabler of the convergence between datasciences and neurosciences, how it has come about and how it is described? Computational neuroscience has emerged in the twenty-first century as an interdisciplinary field that combines neuroscience, computer science and experimental psychology to investigate the principles and mechanisms underlying the brain's information-processing capabilities.[27] It aims to model artificial neural networks and understand how they produce cognitive functions like learning, vision and auditory perception.[28] Computational neuroscientists use research methods developed in datasciences to simulate and analyse neural systems, drawing insights from both neurobiological data and computational ideas. Computational neuroscience is now driven by the conviction that nervous systems can be understood as naturally evolved, analogue and parallel computers. This perspective has led to the development of new mathematical and computational approaches to modelling biological information processing. Methods in computational neuroscience include: coherent infomax (a theory of local neural processors that use contextual field inputs to modulate the transmission of receptive field information, in a cooperative manner that maximises overall coherence); non-linear latent structure analysis (computational learning rules that can discover statistically related variables in input data, without requiring supervision, to uncover the underlying structure); contextually guided computation (algorithms that use contextual information to guide both learning and moment-to-moment processing, inspired by neuroscientific findings on the role of context in the brain); and comparisons to deep learning (while deep learning algorithms have transformed artificial intelligence, context-sensitive cooperative computation may better reflect the way the human mind works, being more closely grounded in neuroscience and psychology).[29] As we noted earlier, computational neuroscience is distinct from experimental neuroscience in that it focuses on developing and testing hypotheses about the functional mechanisms of the brain, often using computer simulations to model the complexity of neural systems. This approach aims to uncover the principles underlying brain function, rather than simply rebuilding the brain in detail on a computer. There is an ongoing tension in the field between incorporating biological details and maintaining model simplicity to illuminate core principles. This tension can be traced to cybernetic conferences and its principal figures such as von Neumann, Wiener and Turing using the concept of 'effective calculability' and the architecture of digital computers as models for information processing. Computational neuroscience increasingly integrates datasciences and neurosciences by developing a new imaginary of the brain and rewriting the imitation game playbook. This convergence is not merely about understanding or intervening in our brains but also remodelling them towards a supposed singularity.[30] William Phillips, for example, insists that modelling, visualisation and learning in computational neuroscience are not simulations but realisations.[31] To understand how the game is no longer imitation but integration or singularity we will need to consider modelling, visualisation and learning to be key methods in computational neuroscience drawn from datasciences.

Modelling

If machine learning has emerged as an interdisciplinary field that enables computers to learn and improve automatically from data, with connections to pattern recognition, data mining, adaptive control, statistical modelling, data analytics and data science, modelling has become the essential method for simulating of the operation and functioning of the imaginary of the brain, so much so that some even believe as we have seen above modelling is not simulation but realisation. Although modelling the brain even in full detail on a computer may not necessarily lead to a better understanding of how it works, it is widely acknowledged that modelling provides insights into certain aspects of brain functions, such as synaptic responses and in turn what can be done with machines.[32] The important point is the limitations in capturing the symbolic processing capabilities of the brain and its separate domains within cognitive science do not inhibit modelling as they enable experimental practices for designing and building learning machines.

Modelling relies on techniques for creating spaces of relations between and among data points, rather than predicting or anticipating its future states of this space of relations. We have seen this shift from deterministic to stochastic and heuristic predicting in datasciences and the techniques based on cluster analysis for drawing out spaces of relations in correspondence analysis. Modelling in computational neuroscience is a species of cluster analysis and correspondence analysis in investigating neural networks as a new imaginary of the central nervous system and perhaps its methods illustrate most vividly the convergence between datasciences and neurosciences.[33]

Visualisation

The methods used for visualising neural circuits and brain connectivity to further neuroscience research have their origins in cluster analysis and correspondence analysis. As an integral element of modelling, visualising produces an imaginary of the central nervous system as an object of realisation. The images that visualisation produce are not representations of the central nervous system as they are not about just mapping anatomical connections (the connectome) and capturing the dynamic electrical activity underlying cognitive processes. There are several large-scale brain mapping initiatives, such as the Human Connectome Project and the 1,000 Functional Connectomes Project, which are using advanced imaging and computational methods to study brain networks (Bielza and Larrañaga 2020, 19–20). However, we can imagine that, with the human brain containing billions of neurons and trillions of synapses, such mapping exercises pose significant challenges in terms of storage and representation. These challenges drive statistical and computational methods in neuroscience and changes the meaning of visualisation from mapping to modelling with iterative parameter estimation, resampling techniques and mathematical statistics. This transforms analytics into a computational discipline visualising the brain as a space of relations. The visualisation of the brain's architecture with information processing encourages the design of more effective systems and algorithms, particularly in the field of machine learning. These developments suggest that the brain's ability to handle novelty, complexity and ambiguity can be used as a model for developing advanced computational systems, as Deutch ([1963] 1966, 86) dreamed.[34] The critical role of visualisation with computational methods for understanding the brain's structure and function drives neuroscience and computer science ever closer (Lytton 2002, 98–9).

The significant developments in recent decades, particularly due to new brain imaging technologies must be seen against this drive for modelling learning machines.[35] While these technologies allow researchers to study the brain's structure, function and activity in unprecedented detail, they are also being used in creating intelligent agents.[36] Take, for example, neuroimaging, which encompasses a range of technologies, including magnetic resonance imaging (MRI), functional MRI (fMRI), positron emission tomography (PET) and electroencephalography (EEG). MRI and fMRI provide high-resolution anatomical and functional information, respectively, enabling the visualisation of brain structures and neural activity associated with cognitive tasks. PET and SPECT (single-photon emission computed tomography) measure changes in blood flow and metabolism, while EEG and MEG (magnetoencephalography) record the electrical and magnetic signals generated by neuronal activity (Bielza and Larrañaga 2020, 19, 30). In addition to these macroscopic techniques, researchers have developed methods to study the brain at the microscopic level. Confocal microscopy and electron microscopy allow for the detailed examination of individual neurons and their structures, while calcium imaging and voltage imaging enable the tracking of neuronal activity patterns. Some techniques are interventionist and experimental: optogenetics and other techniques also allow for the targeted manipulation of neural circuits, providing data into their functions (Bielza and Larrañaga 2020, 30).

We can say with some confidence that the amount of data generated by these neurotechnologies has been foraged by data-driven computational neuroscience. Now statistical and computational models visualise and analyse the vast amounts of information collected from various sources, such as electrophysiological recordings, neuroimaging data and genetic data. These models are then evaluated with experimental verification.[37]

That the combination of advanced imaging techniques and computational approaches has not only enabled neuroscientists to gain a particular understanding of the brain's structure, function and connectivity but also accelerated the drive to build intelligent agents by technologists is significant. As this interdisciplinary field of visualisation as modelling continues to evolve, with ongoing advancements in both experimental and analytical methods, we not only institute an imaginary of the complex workings of the human brain with technologies that imitate it but also accumulate knowledge for producing intelligent agents constituting learning machines. This is the transformation from machine learning into learning machines as a key moment in the birth of sensory power.

Machine learning

The production of learning agents that can act intelligently has become the driver of building learning machines. The central idea is that intelligent behaviour can be designed and imitated through the principles of control, communication and feedback. There are of course many objections and warnings that have been raised against the imitation game and that machine intelligence can reach and surpass human intelligence but the machine intelligence, or at least the debate over it, has been relentlessly intensifying.[38] The crucial point to cut through all this is perhaps to recognise that the meaning of intelligence in learning machines undergoes a transformation. We returned to the origins of the concept to see how it functioned to understand how it changed. Alan Turing may have cautioned against imagining intelligence as an abstract machine.[39] The important point he made in various occasions that human intelligence is recognisable only within certain social spaces and their meaning change from domain to domain and social space to social space has faded.[40] Wiener too may have focused on the processes of steering, communication and self-regulation, and understood the brain and computer as information-processing systems in different and diverse domains of application but that nuanced articulation has dramatically faded.[41]

What has happened instead is that cybernetics has evolved into computational neuroscience and increasingly been driven towards building abstract and concrete learning machines. The production of computational agents that recognise their space with sensory apparatuses and act upon it in a reasonable way, aiming to maximise their performance based on a given objective has become the driving operative principle of developing machine intelligence.[42] The fundamental question that arises from these learning machines is how intelligence is signified and how its performance is measured. The design of learning machines focuses on the physical symbol system development, reasoning and acting, representation schemes, planning horizons, and dealing with uncertainty in sensing and action effects but all these are generated with creating objective to achieve something.[43] To put it differently, although it is presented as developing abstract machines, the drive is towards developing concrete machines capable of performing specific and assigned objectives.

Those involved in the production of learning machines work with concrete machines as 'rational agents' that are designed to achieve the best outcome or expected outcome in a given gameplay. The difference between abstract and concrete machines is similar but not identical to that between general and special intelligence.[44] The idea that there must be 'value alignment' between intelligent agents and their designers indicates that technologists understand agents as concrete machines. The relationship between the intelligent agent and its designers is understood as dependent. The language of intelligent agent design such as the properties of games, whether they are deterministic, episodic or sequential, static or dynamic, and discrete or continuous, is about the design of agents as concrete machines (Russell and Norvig 2022, 60–75). The learning agents are those that improve their performance over time through feedback and adaptation; the agents may not be programmed with already determined algorithms but they will have concrete goals and objectives. This is where we can understand how and why modelling as opposed to programming or predicting becomes key to developing learning agents. These agents are not understood as automata as automated machines may imply; rather, they are imagined as self-organising agents with capacities for initiative and even judgement but with concrete objectives rather than abstract capabilities. Modelling learning machines maybe based on indeterminacy and uncertainty rather than automated action but learning machines learn with concrete goals and objectives. We discussed earlier Turing's (1950) argument that the meaning of learning in learning machines is genuine and fundamentally the same as education that children undergo, which involves random or stochastic events. However, he believed not all learners are equally capable and that some are subcritical, while others are supercritical. Can deep learning solve the problem of creating supercritical intelligence?[45]

That deep learning enables machines to perform operations that were previously only achievable by humans, such as speech recognition and object recognition in images and spaces should not be surprising. This idea developed through research on artificial neural networks inspired by the hierarchical structure and information-processing capabilities of the human brain. The key elements of deep learning include its ability to integrate complex representations from raw data, its scalability to handle large datasets and its flexibility in adapting to diverse operations. Accordingly, deep learning algorithms have been relentlessly applied in a wide range of domains, from speech recognition and natural language processing to computer vision and high-frequency trading. As long as we remember that all these are concrete machines designed to accomplish concrete objectives.

We can now see that the convergence between datasciences and neurosciences has been crucial for machine learning, both by inspiring the architectural design of artificial neural networks and by providing insights into the principles of learning and information processing in natural and artificial. Conversely, the success of deep learning has also led to a particular imaginary of the brain's 'computational' mechanisms, which now guides neuroscientific research.[46] We need to note why the limitations of deep learning are often cast as computational limitations. These include the need for large, labelled datasets, the lack of interpretability and transparency in deep learning models, and the energy-intensive nature of deep learning computations. Enormous energy (both human and non-human) is being expended to overcome these challenges, such as developing more efficient hardware, incorporating contextual information and integrating deep learning with other techniques like reinforcement learning. If indeed machine learning is the latest stage in the imitation game, we are experiencing its outcome as the learning machines.

The invention of the learning machine

The concept of intelligent agents and their design principles is a key element in the construction of learning machines. We can outline the following principles from concepts, practices and methods of the imitation game that invented the learning machine. An agent is an entity that perceives its space through sensors and acts upon that space through actuators.[47] The agent's behaviour is shaped by the agent function, which maps any given sequence of perceptions to an action. A specific concept of rationality dominates thinking about learning machines – a rational agent is one that behaves according to given performance measures, the agent's knowledge, the available actions and the agent's perception sequence.

The design of learning machines determines the properties of gameplay, which can vary along several dimensions such as observability, number of agents, determinism, episodic or sequential nature, static or dynamic states and discrete or continuous actions. These properties determine the appropriate design for the agent. There are basic types of agent designs: simple agents, response agents, purposive agents and decision agents. Simple agents choose actions based solely on the current perception, while response agents maintain an internal state to track aspects of the game that are not apparent in the current perception. Purposive agents choose actions to achieve their goals, while decision agents try to maximise their expected utility and based on diagnostics render decisions. Of interest are always the learning agents, which have a separate learning element that modifies the performance element to improve the agent's behaviour over time. Russell and Norvig (2022, 75) claim that the learning element uses feedback from a critic to update the agent's knowledge, goals and utility function. Who is that critic?

Agents and spaces are inseparable for understanding a particular conception of rationality that shape their design. The overall understanding of an agent as a learning machine that perceives its space through sensors and acts through actuators is at the centre of its rationality. When we say learning machines, we mean machines that are composed of agents that can be human and non-human, hardware and software, playing in a specific game. The autonomic agent embodies all sorts of agents including humans and non-humans. The perception sequence that allows the autonomic agent to perform its functions involves the memory of everything it has perceived. This is precisely the meaning of learning in learning machines: they learn from the actions of other players during their gameplay for the specific game they are designed for as a concrete machine.

The actions of the autonomic agent depend on how the agent functions by mapping perception sequences to actions. And when an agent acts it follows its function described by its design. The function is determined by the position of the agent within a system designed to perform or enact a strategy. So, what constitutes the conduct or behaviour of a learning machine within a system and its rationality? The design of learning machines escapes or evades the question of rights and responsibilities by enacting rationality as performance of the autonomic agent measured against its implicit or explicit design principles. This must be understood as absolutely crucial in the design of intelligence machines: their intelligence embodies and circumscribes intelligence that can only be measured against their design principles.

The design of intelligent agents is therefore possible only with performance measures to ensure agents behave as desired by their designers. These performance measures are based on perceptions and knowledge they accumulate. Thus, the rationality of an intelligent agent does not require omniscience but is measured against expected performance. The intelligent agents cannot reason against their design principles in relation to the game for which they are deployed (if they do this it will be considered a malfunction). If indeed information gathering and learning are essential components of rational agents, the interpretation of these as perceptions, processing them as knowledge and acting on them as intelligent agents requires a special type of decision-making. It is in this sense that the intelligent agents are neither autonomous nor automatic but autonomic.

Through sensors and actuators learning machines not only adapt and adjust to their spaces (automatic) but also thorough feedback mechanisms they can gradually change their spaces (autonomous). How well they perceive their spaces represented by sensors, how well they respond to their actuators and how well they perform actions constitute an autonomic loop where their performance is measured and enhanced. They are autonomic agents that constitute the autonomic agent.

We have seen earlier that Alan Turing recognised how the nature of social space is conceived and perceived by the designers of learning machines and what purposes of their design constitutes their specific intelligence. It is vital to understand that a game is always an operational game. It is never a universal game but a game designed and imagined with a system of strategies and technologies. The operational games are the problems to which rational agents are solutions.[48]

The learning machine operations then involve performance, game, actuators and sensors. The learning machines are the concrete machines of sensory power as they can be designed to function as sensory apparatuses for fully observable, partially observable or unobservable operational games. The decision-making of learning machines consists in sensing and evaluating the perceptions and actuating specific responses as actions.

To put it all differently, in order for intelligent agents to perceive and act in physical or social space, their operation must be precisely specified by designing agents. If, in fact, learning machines improve their performance by learning, their designers learn by the performance of their learning machines. It is this loop that constitutes the principle of the accumulation of autonomic agents and machinic capital as the operating principle of sensory power. The importance of the invention of the learning machines is that it transforms both the subjects of politics and politics of subjects by accelerating the creation of polities as learning machines – an acceleration that radically transforms the citizenship game invented 2,500 years ago.

Notes

1. A contemporary account by Neil Watson and Marc Breedlove, *The Mind's Machine* ([2012] 2021), takes the brain–machine analogy as given, needing no explanation. When they discuss what they call a longstanding

'metaphor' of the brain as a machine, they nonetheless think modern neuroscience aims to understand how the brain's physical operation gives rise to the mind and its various cognitive processes. Their account of localisation of brain's functions in different brain regions and how the development of modern brain mapping techniques contributed to this image remains practically silent about the brain–machine metaphor.

2. Daniel Smail (2008) presents a deep history of humans that centres on biology, brain and behaviour, challenging the dominant historical narrative that begins with the rise of civilisation. The claim is that the discipline of history has long avoided geological time, fashioning a view of history that excludes the Palaeolithic and Neolithic eras. Instead, Smail proposes a narrative that traces the biological legacies of the deep past into the present, highlighting how features of culture are wired into human physiology. He explores the role of the brain in negotiating the complexities of human social living, and the influence of neurochemicals, moods, emotions and feelings that have a deep evolutionary history. Smail sees the plasticity of these brain–body states, arguing that they do not dictate behaviour but rather provide a backdrop against which people make decisions and take actions. He presents this view as a critique of evolutionary psychology, which has been associated with theories of innate and hereditary difference.
3. 'Instead, the large human brain evolved over the past 1.7 million years to allow individuals to negotiate the escalating complexities posed by human social living' (Smail 2008, 113).
4. 'The neurohistorical approach embraces the recent interest in cognition as well as all histories that emphasise noncognitive aspects of the brain, including the history of emotions. But to acknowledge the physical or neural reality of moods and predispositions is not to adopt a crude genetic determinism. Even less does it invoke the illusory search for an essential human nature engineered by natural selection in the distant past, a stance that can only lead to a simpleminded history-without-change' (Smail 2008, 114).
5. There is an immense literature on the brain and we will have opportunities to discuss this in this chapter. It is possible to group the literature into original, interpretive, critical and popular. Of course, many authors straddle interpretive, critical and popular positions but our attention is mostly on the original literature developed on site with practices.
6. 'The brain exhausts itself in its phenomena. It exists, of course, as an organ. But, critically speaking, it doesn't have an archepresence outside of its different exteriorisations' (Malabou 2022, 311).

7. 'In hardware, neuromorphic computing takes its inspiration from observations of the complexity of the biological brain and considers brain knowledge as principles that can be applied to the design of hardware engineering systems. Seminal work on bio-inspired microelectronics culminated in the book published by Mead (1989). More recent examples of large-scale neuromorphic systems include (a) the IBM TrueNorth chip, which is the outcome of a decade of work by the DARPA SYNAPSE program aimed at delivering a very dense, energy-efficient platform capable of supporting a range of cognitive applications, (b) Neurogrid, which is based on the heritage of Mead (1989), and uses subthreshold analogue circuits to model neuron and synapse dynamics in biological real time, (c) the SpiNNAker project, which has developed a massively parallel digital computer whose communication infrastructure is motivated by the objective of modelling large-scale spiking neural networks with connectivity similar to the brain in biological real time, and (d) the BrainScaleS neuromorphic system developed at the University of Heidelberg with the aim of implementing physical models of neuronal processes' (Bielza and Larrañaga 2020, 30).
8. Malik et al. (2023); Trappenberg (2022); Bielza and Larrañaga (2020); Churchland and Sejnowski ([1992] 2017); Bower (2013); Lytton (2002).
9. Churchland and Sejnowski ([1992] 2017); Bower (2013); Changeux and Chavaillon (1996); Ray (2024).
10. Alan Turing is a major figure of the convergence between datasciences and neurosciences. Although his interest in neurosciences was limited, he was well aware of the history of the brain–machine analogy in both the UK and USA. His most important writings are compiled in Turing (2004), although we will mostly refer to the original publications (1936; 1950). We will also make considerable use of the literature on Turing, especially Bernhardt (2016), Dyson (2012), Warwick and Shah (2016) and Gonçalves (2023).
11. Almost a century before Turing, Ada Lovelace (1843) thought about whether Charles Babbage's (1791–1871) machine could do the kinds of things humans could do.
12. Margaret Boden (2016, 9) notes that Turing had written a fuller version of this article after the war but the Official Secrets Act prevented its publication.
13. Turing expresses this cautious view quite well: 'One way of setting about our task of building a "thinking machine" would be to take a man as a whole and to try to replace all the parts of him by machinery. He would include television cameras, microphones, loudspeakers, wheels

and "handling servomechanisms" as well as some sort of "electronic brain." This would of course be a tremendous undertaking. The object if produced by present techniques would be of immense size, even if the "brain" part were stationary and controlled the body from a distance. In order that the machine should have a chance of finding things out for itself it should be allowed to roam the countryside, and the danger to the ordinary citizen would be serious. Moreover, even when the facilities mentioned above were provided, the creature would still have no contact with food, sex, sport and many other things of interest to the human being. Thus although this method is probably the 'sure' way of producing a thinking machine it seems to be altogether too slow and impracticable' ([1948] 2004, 420).

14. Dyson (2012, ix) says that 'Von Neumann's project was the physical realization of Alan Turing's Universal Machine, a theoretical construct invented in 1936. It was not the first computer. It was not even the second or third computer. It was, however, among the first computers to make full use of a high-speed random-access storage matrix and became the machine whose coding was most widely replicated and whose logical architecture was most widely reproduced. The stored-program computer, as conceived by Alan Turing and delivered by John von Neumann, broke the distinction between numbers that mean things and numbers that do things. Our universe would never be the same.'
15. Conway and Siegelman (2005, 252). There is a fascinating history of the involvement of both von Neumann and Wiener in the development of the atomic bomb but it is beyond our scope (Heims 1980; Conway and Siegelman 2005).
16. Several scholars have provided accounts of the subsequent development of the brain–machine analogy and how both became increasingly sophisticated. An early scholar was Hubert Dreyfus (1972; [1979] 1992; 1986), who focused almost entirely on the question of whether machines can think like humans. Margaret Boden (2006; 2016) also focuses on that question but with more attention to the recent development on intelligent agents.
17. Churchland and Sejnowski ([1992] 2017, 7) express this more emphatically: 'nervous systems and probably parts of nervous systems are themselves naturally evolved computers – organically constituted, analog in representation, and parallel in their processing architecture.'
18. Churchland and Sejnowski's ([1992] 2017) *The Computational Brain* is a pioneer of the convergence between datasciences and neurosciences.

19. Ray Kurzweil (2006), who popularised the concept of singularity, which means the convergences between the brain and the machine, attributes it to von Neumann in his foreword (Kurzweil 2012) to the new edition of von Neuman's *Brain and Computer* ([1958] 2012).
20. Poole and Mackworth ([2010] 2023, 19) put this clearly: 'Modularity is the extent to which a system can be decomposed into interacting modules that can be understood separately. Modularity is important for reducing complexity. It is apparent in the structure of the brain, serves as a foundation of computer science, and is an important part of any large organisation.'
21. Smail (2008, 139) describes this as follows: 'These cognitive modules are like computer programs. In theory, if one could map the neural networks of the brain, one could actually see these programs in the form of vast, interconnecting sets of neurons. Here is the evolutionary or historical component: these modules were engineered by the process of natural selection acting on humans in the ancestral environment. Some modules, like basic fears, urges, and other predispositions, are identical to those found in primate or mammalian brains and indeed derived from them. Other modules, like deep grammar, emerged more recently and are unique to humans. According to evolutionary psychologists, all have remained largely unchanged since the origin of the species some 140,000 years ago.'
22. Bielza and Larrañaga (2020, 5) explain that the brain is decomposed into subsystems or modules and that 'anatomical modules in the spatial domain are present in cortical mini-columns or columns, whereas short- and long-term memory are the modules in the temporal domain. Within-module elements are more highly connected than between-module elements.'
23. 'A network is simply a group of neurons that become active under certain conditions. Networks are the way in which basic human processes such as learning, memory, thinking, planning, feeling, and decision-making take place' (Ray 2024, 35). 'How are neurons connected in a network? The answer may seem strange. Neurons are neither totally random in their connections with other neurons nor totally patterned. It appears that neurons are connected to one another in the same way that all humans on this planet are socially connected' (Ray 2024, 45). 'Neurons have numerous short-distance local connections which taken together can be considered as a hub or module. From these hubs there are more long-distance connections to other hubs. Local hubs can be made up of neurons which connect with each other over very short distances. Such

connections are seen in gray matter. Gray matter appears dark and can be seen in the outer shell of the brain. Gray matter contains the synapses, dendrites, cell bodies, and local axons of neurons' (Ray 2024, 45).

24 'Networks with a specific architecture and specialised information-processing capabilities are composed into larger structures that are able to perform even more complex information-processing tasks. System-level models are important in understanding higher-order brain functions' (Trappenberg 2022, 6). 'The nervous system is distributed throughout the whole body. Some of the peripheral nervous system include sensors such as touch sensors or sensors for auditory signals. Some of those sensors like the eyes are in themselves already highly sophisticated neural systems, and the brainstem already processes sensory signals to produce fast responses such as reflexes' (Trappenberg 2022, 7).

25. 'Neurons are the basic structural components of the brain. A neuron is an individual cell, specialised by architectural features that enable fast changes of voltage across its membrane as well as voltage changes in neighbouring neurons. Brains are assemblies of just such cells, and while an individual neuron does not see or reason or remember, brains regularly do. How do you get from ion movement across cell membranes to memory or perception in brains. What is the nature of neuron-neuron connectivity and interactivity What makes a clump of neurons a nervous system?' (Churchland and Sejnowski [1992] 2017, 1).

26. As Malabou ([2004] 2008, 5) says, 'talking about the plasticity of the brain thus amounts to thinking of the brain as something modifiable, "formable," and formative at the same time. Brain plasticity operates, as we shall see, on three levels: (1) the modelling of neuronal connections (developmental plasticity in the embryo and the child); (2) the modification of neuronal connections (the plasticity of synaptic modulation throughout life); and (3) the capacity for repair (postlesional plasticity).'

27. Lytton (2002); Trappenberg (2022); Malik et al. (2023); Bielza and Larrañaga (2020).

28. 'The expression "computational" in computational neuroscience reflects the role of the computer as a research tool in modelling complex systems such as networks, ganglia, and brains. Using the word in that sense, one could have also computational astronomy or computational geology. In the present context, however, the word's primary force is its descriptive connotation, which here betokens the deep-seated conviction that what is being modelled by a computer is itself a kind of computer, albeit one quite unlike the serial, digital machines on which computer science cut its teeth. That is, nervous systems and probably parts of nervous systems

are themselves naturally evolved computers-organically constituted, analogue in representation, and parallel in their processing architecture' (Churchland and Sejnowski [1992] 2017, 7). 'Computational neuroscience is the theoretical study of the brain used to uncover the principles and mechanisms that guide the development, organization, information processing and mental abilities of the nervous system' (Trappenberg 2022, 3).

29. William Phillips (2023) interprets the recent developments in neuroscience with computational neuroscience with the label 'the cooperative neuron'. Phillips claims that the conceptual revolution occurring in neuroscience in the understanding of the cellular foundations of mental life and its disorders has important consequences for methods. For him conscious perception, thought and action and the recently discovered capabilities of pyramidal cells, a special class of neocortical neurons widely distributed through the cerebral neocortex, show that neurons work together rather than compete. Phillips says that many of these neurons are sensitive to context, with two functionally distinct sets of inputs – one that provides the information they usually transmit, and another that can selectively amplify that transmission when useful in the current context. This context-sensitive, cooperative information processing simplifies the production of coherent percepts, thoughts and actions related to current goals and needs. What this interpretation shows from our perspective is how the understanding of brain is now entwined with how we understand the machine.
30. As Phillips argues, 'when theories of mental function loosely inspired by neurobiology became prominent in the 1980s under the titles of parallel distributed processing (PDP), or connectionism, they raised doubts in the artificial intelligence (AI) community because of what were thought to be inherent limitations in their ability to learn. The tables are now fully turned by the undeniable success of neurobiologically inspired machine learning algorithms, such as those of deep learning. Many researchers now think that the capabilities of deep learning algorithms may guide us towards a better understanding of learning in the neocortex' (Phillips 2023, 274).
31. As Phillips insists, 'the abstract neural nets used to explore the capabilities of the style of neuronal computation that we call "coherent infomax" are realizations of that style of computation – they are not simulations of it' (Phillips 2023, 258).
32. Trappenberg (2022, 20), for example, acknowledges that 'modelling the brain is not a contest in recreating the brain in all its details on a

computer. It is questionable how much more we would comprehend the functionality of the brain with such complex and detailed models.'

33. The literature on computational brain is extensive. See Miller (2018), Churchland and Sejnowski ([1992] 2017), Lytton (2002), Sterratt (2011) and Trappenberg (2022).
34. Bielza and Larrañaga (2020, 30) outline these possibilities.
35. As Bielza and Larrañaga (2020, 30) argue, 'The brain is able to work with novelty, complexity, and ambiguity, whereas computers are fast and very precise. However, the knowledge about the brain can serve as inspiration in the design of machines, both software and hardware, for example by applying the lessons learned from connectome graphs to making computers smarter. In software, strategies used to unravel the algorithmic specializations of the sensory cerebral cortex are inspiring the next generation of high-performance machine learning.'
36. As Bielza and Larrañaga (2020, 22) say, 'the progress in computer technology led to changes in statistical methodology and statistical ideas. Examples include the inversion of large matrices, iterative methods for the estimation of parameters, and the introduction of various types of resampling methods, such as bootstrapping, dynamic interactive visualization of data sets, and Bayesian statistics, a field that would not have been developed without the assistance of computers. Statistics was regarded as a mathematical discipline; however, it is currently regarded as a computational discipline focused on understanding the scientific problem and providing a correct interpretation.'
37. For further examples see Bower (2013) and Sterratt (2011).
38. The literature is now fairly extensive but especially on how to think about the key ethical and social issue of artificial intelligence see Risse (2023), Possati (2022), Coeckelbergh (2022), Natale (2021), Dyer-Witheford, Kjøsen and Steinhoff (2019) and Wilson (2010).
39. See Dyson (2012), Bernhardt (2016) and Gonçalves (2023).
40. Bernardo Gonçalves (2023) provides a convincing interpretation of Turing as an ironic thinker. He identifies three distinct images of Turing: a utopian and radically progressive thinker, an infantilised and politely humorous genius, and an irresponsible scientist reminiscent of Frankenstein. Gonçalves develop the image of an ironic thinker arguing that Turing's irony may have been lost on some of his followers. For Gonçalves there is enough to indicate that by using irony Turing attempted to shift the focus from electronic brains used for warfare to studies of nature and mathematics.
41. Conway and Siegelman (2005) provide this view of Wiener.

42. Russell and Norvig (2022, 548).
43. Russell and Norvig (2022) is the most comprehensive introduction to the principles of machine intelligence from the perspective of the convinced.
44. For general and specific intelligence see Boden (2006, 149ff, 548ff). See Deleuze and Guattari ([1980] 1987) for abstract and concrete machines.
45. Phillips (2023, 274), for example, observes that, 'when theories of mental function loosely inspired by neurobiology became prominent in the 1980s under the titles of parallel distributed processing (PDP), or connectionism, they raised doubts in the artificial intelligence (AI) community because of what were thought to be inherent limitations in their ability to learn. The tables are now fully turned by the undeniable success of neurobiologically inspired machine learning algorithms, such as those of deep learning. Many researchers now think that the capabilities of deep learning algorithms may guide us towards a better understanding of learning in the neocortex.'
46. Sejnowski (2018, 25) observes that 'the Deep Learning Revolution has two intertwined themes: how human intelligence evolved and how artificial intelligence is evolving. The big difference between the two kinds of intelligence is that it took human intelligence many millions of years to evolve, but artificial intelligence is evolving on a trajectory measured in decades.'
47. We are following Russell and Norvig's (2022, 54–78) description of intelligent agents as the basis of our discussion here.
48. Bruno Latour (2004) provides a fascinating discussion of Alan Turing (1950), who distinguishes between subcritical and supercritical learning machines. For Turing, a learning machine can be considered intelligent only if it develops supercritical capabilities, setting out different goals and objectives beyond its design constraints. While Latour's primary objective is to reevaluate the concept of critique in social and political theory, it is equally applicable to the study of learning machines.

4

The autopoietic subject

An historical ontology of datasciences and neurosciences reveals their convergence as the force that is shaping the imaginary of learning polities and its subjects and objects in the twenty-first century. The enactment of this imaginary combining neurosciences and datasciences shows how an autonomic subject is produced by sensory power. To be direct, the subject of sensory power is neither autonomous nor automatic but it is autonomic. This is the operative principle of the imaginary of polities as learning machines. An imaginary of polities that embodies machine learning governs with the operative concepts, practices and methods of datasciences and neurosciences and produces subjects and objects that are neither automatic, because they have agentic capacities, nor autonomous, because their agentic capacities are governed by sensory power.

Unlike anything we have known previously, polities as learning machines are without territory, population or wealth. This is not naively to announce the death of polities with territory, population or wealth but to recognise the emergence of new kinds of polity as learning machines that are nestled within but irreducible to other polities and that are accumulating news forms of capital and subject peoples. Governing learning machines as transient, transversal and revaluating polities and their autonomic subjects generates logics, affects, effects and strategies and technologies of power. If the autonomic subject is both the affect and effect of sensory power, understanding the strategies and technologies of sensory power and forms of capital that they accumulate becomes imperative. To paraphrase Foucault with Bourdieu, if accumulation of different forms of capital would have been impossible without accumulating different forms of subjects, learning machines would have been inoperable without sensory power.

If power as such and power as performatives as doing things with numbers and neurons invent a new form of power – sensory power – the autonomic subject is its most relevant affect and effect. If the autonomic subject of learning machines is neither automatic nor autonomous but autonomic, and if polities are neither territorial nor permanent but transversal and remanent, it is imperative to understand the concrete situations in which sensory power produces the autonomic subject. The autonomic subject is not divided by lines and is not a multitude. It is combinative because its integration into a multiplicity of polities as learning machines generates its autonomic qualities. The autonomic subject is at once abstract and concrete, since it is neither universal nor particular, but transversal. The autonomic subject is abstract because it functions exactly the same way and everywhere. Yet it is combinative since each abstract autonomic subject is composed of a unique assemblage of connections to polities and this combination is dynamic and fluid. The combinatory of the autonomic subject is the spatially and temporally dynamic assemblage of its connections to multiple polities and their logics, affects and effects. The autonomic subject becomes aware of things and peoples through a unique sensory apparatus it and only it possesses. Yet its sensory apparatus is an assemblage of machines, sensors, devices and bodies. How does the sensory apparatus generate this autonomic subject and how should we name it?

The citizenship game

To understand the citizenship game as it is played by sensory power in the city, it is necessary to outline a history of the city–citizenship relationship. The origins of the city always haunt writing about polities and politics. Over the last 200 years the dominant imaginary has been that for long periods of time humans around the world appeared to change little: they lived in small tribes as hunters and gatherers. Then, around 12,000 years ago, the pace of change began to accelerate. Over the next few millennia, humans in the 'fertile crescent' of the Middle East, the foothills of the Indus Valley and the valleys of the Nile and Yellow Rivers, and some centuries later in valleys and mountains of Mesoamerica, developed the ability to produce their own food by cultivating land and domesticating animals. As these humans became more sedentary, the ability to produce their own food led to the ability to accumulate surplus. Then, in the foothills east of the Tigris, a revolution – an urban revolution – took place about 6,000 years ago: humans 'settled together' – as the ancient Greek narrative would have it, synoecism (or sedentism) – and the city was born.

Or, alternatively, this is how in the nineteenth and early twentieth centuries the idea of the city was invented. The question of the origins of cities held the greatest fascination in this period about how the 'city states' in the great river valleys of the Tigris and Euphrates, the Nile, the Indus and the Yellow Rivers emerged about 6,000 years ago. The origins of these early cities became even more intriguing when decades of interpretive struggles suggested that, in each of these valley locations, the impetus that led human to 'settle together' in ever larger, denser and more differentiated settlements was independent from each other. As economy and society became more complex, production and exchange of goods (if not services), initially tenuous, began to establish links between them. The emergence of the earliest Chinese polities in the Yellow River valley was interpreted as sufficiently isolated from developments in the Near East and India and that all these early cities were considered original. Moreover, although the ancient Mesoamerican cities flourished much later, from 2,000 to 500 years ago, the Maya, Teotihuacán, Aztecs, Toltecs and ancient Incas were all considered to have developed polities independently of each other. The question that arose about the origins of these city was how their independent rise was possible.

The narratives that the city emerged from an agrarian revolution or that the city itself catalysed an agrarian explosion were deconstructed in the late twentieth and early twenty-first centuries. It became evident that the underlying thread that held these narratives together was not only synoecism (or sedentism), which portrayed cities as aggregations or settlements of people but also orientalism (and Eurocentrism), asserting that the city was most politically developed in Europe and subsequently spread to other world regions – the Greco-Roman hypothesis. To put it bluntly, the driver of the studies on the origins of cities turned out to be a Eurocentric thesis that the citizenship game was invented in European cities.

By the end of the twentieth century these narratives were deconstructed by a generation of scholars and as Eurocentric or orientalist perspectives faded into the twenty-first century. Our young century now interprets the origin of the city as part of the diverse polities that emerged as a consequence of human movements that began almost 100,000 years ago from Africa to Europe, Asia, Australia and the Americas. The birth of the city in human history about 6,000 years ago now appears a part of these movements.[1] This has significant consequences for reworking a long history of power.

It was Deleuze and Guattari ([1980] 1987), inspired by Mumford ([1966] 1967; [1964] 1970), who noted that histories of polities and their government have always been written from the perspective of sedentary peoples seeing nomadic peoples as barbarians. The recent contribution of James Scott (2017) to our understanding of the origins of the state as a sedentary polity written from the perspective of nomadic peoples has been, in many ways, a major development of this insight by Deleuze and Guattari.

Scott (2017) playfully but seriously designates nomadic peoples generically as barbarians and provides an account of the domestication of nomadic peoples into sedentary peoples from about 12,000 years ago to about 6,000 years ago. Scott argues that the emergence of polities (Scott does not distinguish between cities, states and empires and uses 'states' for all early polities) in fertile valleys across the planet earth about 6,000 years ago cannot be merely attributed to the invention of cultivation, as traditional histories written by sedentary peoples would have it. The invention of the polity was preceded by cultivation by thousands of years and sedentism did not directly or necessarily lead to the formation of polities. Scott challenges the narrative that agriculture was a revolution in food production, freeing people to engage in other pursuits for settling in states. Nor does he suggest that early states provided anything of the kind implied by civilisation. On the contrary, early polities were apparatuses of capture of subject peoples by slavery, bondage and war. Scott shows how nomadic peoples outside or at edges of the polity were seen as barbarians (2017, xii). By bringing sedentary peoples into political thinking about the state specifically and all polities (cities, states, empires), Scott proposes that the domestication of animals and domestication of peoples by polities were related: slaves, subjects and the subjugation of women were effects of such domestication.[2]

Scott's reversal of the conventional history of the polity, told by generations of scholars over the past 200 years or so as a progression from nomadism to sedentism, can be used in a number of ways. First, it is an interpretation of history from the perspective of movement rather than settlement. His reversal of sedentism to nomadism serves this purpose. As Scott rightly insists, our knowledge of history is vastly overpowered by sedentism because it is sedentism that stores, archives, collates, assembles, interprets and transmits history from its perspective. Sedentism has accumulated technologies of archival and transmission that are antithetical to nomadism. The peoples on the move or at the edges or outside the polity have not accumulated a similarly durable knowledge and infra- and superstructures, for obvious reasons. Organised polities function as apparatuses for capturing subject peoples and for accumulating surplus for the maintenance and expansion of territories and influence. The apparatuses of government that organise polities rework the domestication of peoples by peoples and all its variations: slavery, debt bondage, forced labour, forced settlement, forced displacement and war. Second, these polities were always fragile government apparatuses. They were never all-too-powerful apparatuses and they often and violently collapsed. Yet these collapses have always been interpreted from the perspective of sedentism as catastrophic. Scott suggests that from the perspective of nomadism periods of these collapses could also have been evidence of emancipation. Finally, sedentism and nomadism are not mutually exclusive, isolated and independent movements. They depend on each other and produce effects through each other. Sedentism produces nomadism and nomadism consumes sedentism. Or, as Deleuze and Guattari say, nomadism and sedentism are immanent and traverse each other. Nomadism is a movement, a becoming that affects sedentism, just as sedentism is a stoppage that captures nomadism (Deleuze and Guattari [1980] 1987, 430).

There is no doubt that the question of the origins of the state as articulated by Mumford, Deleuze and Guattari and Scott challenges historians.[3] What Scott does not recognise, however, is the birth of the kingship game. Although Scott focuses on the early history of states from 12,000 to 6,000 years ago, it is important to recognise the emergence of kingship that invented sovereign power was a key event. From about 6,000 to 2,500 years ago, early states were primarily ruled by the kingship game and the cities as their seats of sovereign power reflected this. The invention of the citizenship game 2,500 years ago was against the kingship game but also its transformation.

Today it is impossible to think about the city as a polity and its relations to other polities without deconstructing this grand narrative. While Mumford, Deleuze and Guattari and Scott have given us different perspectives on the city, it is Weber ([1896] 1998) who most convincingly distinguishes between the kingship game and citizenship game in the long history of power, although succumbing to synoecism and orientalism in this interpretation of the citizenship game (Weber 1921). If we interpret the origins and development of polities not only as agglomeration, settlement or association but as machines in the sense they organise peoples with different forms of power, we can see that the city as a difference machine makes the citizenship game possible.[4]

The sovereign power establishes an imaginary of the city as a polity where kings and god-kings exercise sovereign power and depend on the subservience and obedience of their subjects. The kingship game that persisted from about 6,000 to 2,500 years ago has given us the most enduring imaginary of power, sovereign power. Mumford was astute to attempt to give an account of the city from the point of view of power or more precisely different forms of power. If the ancient city depended on kingship and the personification of sovereign power, the most significant revolution was when the sovereign power was challenged by first an emerging aristocratic class and second a military class. This was the invention of the citizenship game: the introduction into polities a new subject – the citizen-subject – whose position in the city was not authorised by sovereign power but made possible by a combination of sovereign and disciplinary power.

The citizenship game implicates subjects in strategies and technologies of power to shape the conduct of each other and the polity as an arrangement that both facilitates and protects the game. When forces (social classes, social groups) succeed in inculcating their own virtues as dominant, citizenship is constituted as an expression and embodiment of those virtues against others who lack them. To put it another way, citizenship emerges from the point of view of the dominant, which constitutes itself as a universal point of view – the point of view of those who dominate the city and who have constituted their point of view as universal by representing the city as a unity. But the strategies of domination are not universal and have differed widely among various groups. Becoming a citizen-subject is that act where the givenness of the dominant virtues is called into question and their arbitrariness revealed. These acts redefine the meaning and functions of being political by devel oping symbolic, social, cultural and economic practices that citizen-subjects constitute as political, taking different positions in the social space than those that they were previously positioned or assigned.

The citizenship game, as the most enduring challenge to the kingship game, introduces struggles between the superior versus inferior, high versus low, black versus white, noble versus base, good versus evil; it is reversed, transvalued, redefined and the terms, ways and meaning of being citizen-subject are resignified. Becoming citizen-subjects is that moment when freedom becomes responsibility and obligation becomes a right and involves arduous work upon oneself and upon others, building solidarity and alterity simultaneously. If all domination is arbitrary, its affects and effects conceal this. The citizenship game was invented when the dominated, stigmatised, oppressed, marginalised and disfranchised people expose the arbitrary they realise themselves as groups and constitute themselves as citizen-subjects. Becoming a citizen-subject involves a people constituting themselves as agents capable of judgement about just and unjust, taking responsibility for that judgement and associating themselves with or against others in fulfilling that responsibility. Becoming citizen-subjects is that moment when a people develop a mastery of strategies and technologies of power that implicate them in certain ways of acting and begin to transform these ways in different directions than their intended aims and trajectories. Coming full circle, the citizenship game introduces a subject into history as an active force. However, as we have seen earlier, Deleuze ([1962] 1985, 57) argues that it can also become a reactive force if it exercises power to separate its other from what it can do.

The history of the citizenship game has often been written by dominant groups who articulated their identity as citizens and regarded strangers, outsiders and aliens as those others who lacked the properties they instituted as essential for citizenship. (They were once actives forces but became reactive forces.) The dominant views on citizenship often derive not from those who questioned and attempted to overturn its values (active forces) but from those who became its benefactors and inheritors (reactive forces). The challenge is not to reinterpret the history of citizenship game from the viewpoint of the reactive forces, whether they are the oppressed or oppressors, but from the point of view of active forces reversing and transvaluating ways and functions of being citizen-subjects.

To put all this bluntly, if the polity is a difference machine, which simultaneously assembles and disperses various groups and their differentiations across social and physical space and enables them to govern themselves and others by using various solidaristic, agonistic and alienating strategies, any group that achieves domination in the city also claims rights to the city. This is a solidaristic and agonistic strategy that enables dominant groups to constitute themselves as virtuous, righteous and ethical agents, justified in imposing their vision of the citizenship game on other groups. That their vision never remains uncontested in physical or social space enables various strategies of becoming citizen-subjects.

The citizenship game fundamentally changed in the late eighteenth and early nineteenth centuries. Becoming citizens-subjects can perhaps be defined as acts of transfiguration and transvaluation by non-citizens. We may well owe the invention of citizenship game not to citizens but to strangers, outsiders and aliens. In their inventive practices of transvaluation, declaring certain ways of being as intolerable, and questioning their relationship to themselves and others, non-citizens may have invented the citizen game. When kings managed to impose their vision of the world, they generally aimed to end politics.

This is no different in the twenty-first century; however, the dominant imaginary of the city is named: the computable city, the cryptographic city, the cybercity, the digital city, the information city, the intelligent city, the networked city, the platform city, the sentient city, the smart city and the wired city. The polity is a difference machine, and how sensory power reorganises and rearranges the constitutive forces, including sovereign, disciplinary and regulative forms of power, is the object of our analysis. We need to examine the new imaginary of the city as a realisation of a difference machine as constituted by concepts (autopoietic, sentient, platform, networked), practices (capturing, detecting, monitoring, pulsing) and methods (sensing, identifying,

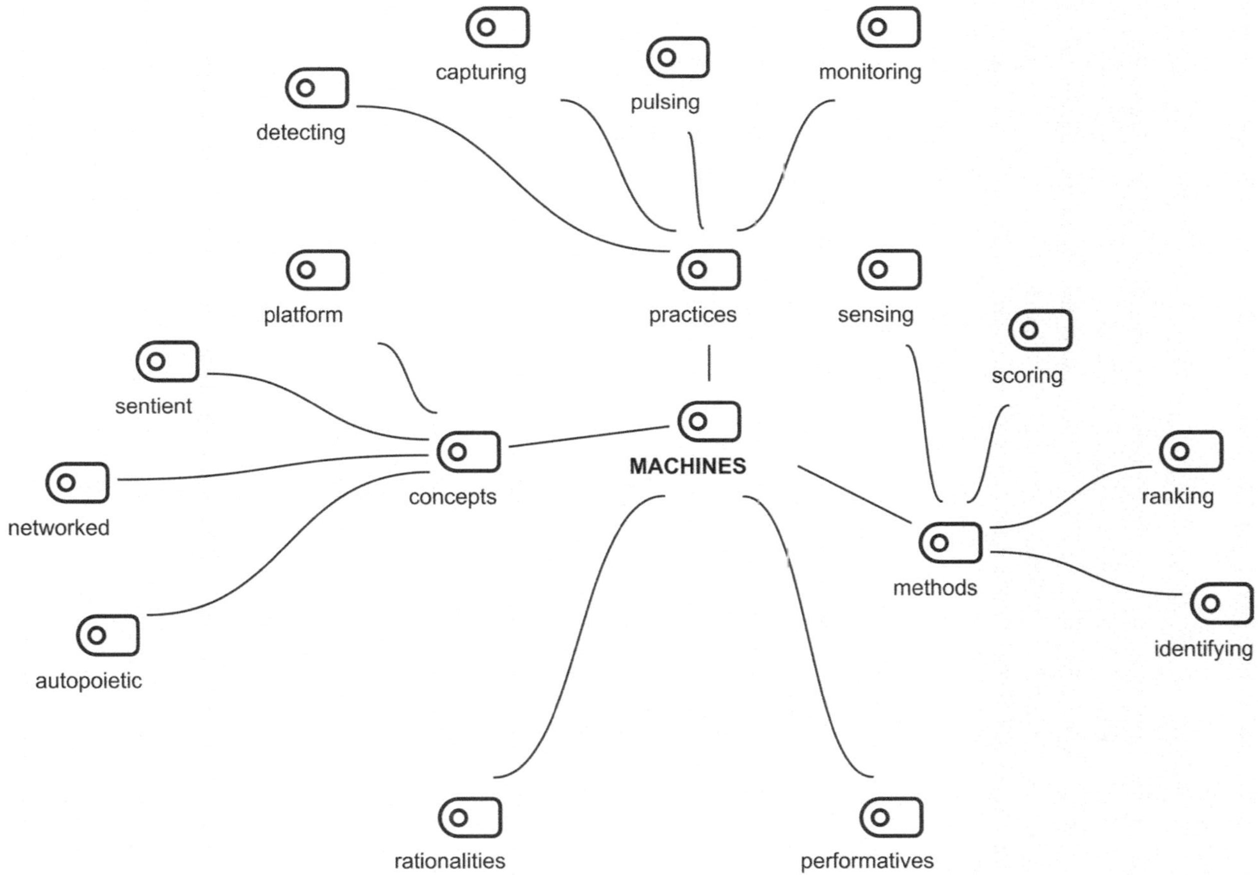

Figure 4.1 Learning machines, polities and transversal clusters

ranking and evaluating). It is these elements that constitute the polity as a difference machine in the twenty-first century, and the creation of the city as a difference machine performs the logics of government embodied in learning machines.

If the primary effects and affects of sensory power are the learning machines, how these machines are assembled into systems by which cities govern and regulate themselves as polities to produce the autonomic subject is primary object of analysis. We have suggested above that the autonomic subject is are neither automatic, because they have agentic capacities, nor autonomous, because their agentic capacities are governed by sensory power. We are going to call it the autopoietic subject.

The scale of the production, implementation and maintenance of the learning machines around the world is staggering. Yes, the experiments are widespread, involving hundreds of cities on every continent – Asia, Africa, North America, South America, Antarctica, Europe and Australia. The discourse on 'smart cities', which clearly dominates over other imaginaries, as a vast and extensive discourse on every continent, produces the autopoietic subject by instituting the city as a learning machine. If we divide this discourse according to their approach into critical, interpretive, original and practical, the discourse on the subject is immense. However, our objective is not a typology or a taxonomy but to identify the essential and decisive operating principles of the actual, proposed or planned smart cities and how their concepts, practices and methods institute this imaginary of the twenty-first century and the autopoietic subject.

Concepts: autopoietic, sentient, platform, networked

The articulation of the city with technological transformations, particularly the rise of intelligent agents and networked systems, has already led scholars to speak of the transformation of the city as a machine into the cybercity. For example, Boyer ([1983] 1994) shows how the professions of architecture, design and planning had already transformed the imaginary of the city as a machine in the nineteenth century by reorganising and rearranging the apparatuses of construction and building. Boyer argues that the disciplinary spaces of enclosure, such as factories, schools and prisons, were rearticulated as machines for modulating control, facilitated by computer technology in cybercities in the twentieth century. If Boyer ([1983] 1994) gives an original interpretation of the city in the twentieth century, Boyer (1996) anticipates the twenty-first-century imaginary of the city. She explores the transformation from the machine to the computer, and the implications for architecture, design and experience. The machinic city of modernism embodied mechanisms of discipline and architectural spaces of enclosure, such as the asylum, prison, factory and school, which served to control and normalise individuals. However, these disciplinary spaces were now dissolved and replaced by a space of flows defined by global computer networks and modulating control. Cyberspace has become a new electronic, invisible space that substitutes for urban experience, where perception of space is increasingly dependent on simulated, predigested digital information. This has led to a fragmented, disappearing city, where reality is no longer used as a referent and space and time take on multiple, non-linear profiles. Consequently, there is a blurring between experience and manipulated media images, with the latter becoming substitutes for direct experience and a recursive structure emerging between real time and deferred or replayed time.

Since Boyer (1996) we have seen the transformation of cybercities into new forms of cities as intelligent systems with autonomous cars, data infrastructures and eventually to projects such as city brains, designing and redesigning cities as self-regulating and self-governing systems that can sense and regulate across an entire city and indeed a network of cities. Shannon Mattern (2021) reacts to this imaginary of the city by drawing attention to the dangers of imagining the city as a computer. Mattern notes that the increasing saturation of simulated and virtual spaces and images has blurred the distinction between experience and reality in cities. If indeed the city as we know it is disappearing and is being fragmented, with space and time taking on multiple, non-linear profiles, whether this is leading to sustainable, equitable, and liveable cities or towards a nightmare scenario of a city operated for and by intelligent agents remains to be seen.[5] What is becoming clear is that the image of the citizen, not only as an inhabitant of the city (settlements) but also as a constitutive polity that governs itself (association), is radically changing with the incorporation of the citizen into the self-organising learning machines. The meaning and function of the citizen in cities as learning machines has changed. If cities have indeed become programmable physical and social spaces in which the image of the citizen is an autonomic subject of a computable, self-monitoring and self-regulating system, what becomes of the meaning and function of being citizen-subjects? This question, which is articulated by Jennifer Gabrys (2014), seems more imperative than the question of whether and to what extent these transformations empower or disempower citizens in government and decision-making.

There is a distinction between the *rationalities* deployed to animate, mobilise and realise various projects to implement and operationalise cities as learning machines, or rather cities as systems of learning machines such as sustainability, equality, inclusivity and efficiency, and *performatives* such as the operative concepts, practices and methods that produce affects and effects. Much of the discourse on smart cities in the twenty-first century are rationalities as justification for developing automated systems to govern them.[6] Massive projects such as India's 100 smart cities project embodies this tension between rationalities and performatives. The governing authorities repeat, cite, iterate diverse rationalities to justify both the investment and arrangement for these projects while simultaneously they also produce performative affects and effects such as technocratic imaginary of 'smart citizenship'. As Ayona Datta (2018) shows, the process occurs through three key elements: enumerations, which are coercions by the state of an urban population hidden from analogue governance technologies; articulations, which are the performances of smart citizenship across digital and material domains that extend historic social inequalities; and breaches, which are the ruptures of the impenetrable technocratic walls around the global smart city, providing a window into alternative and possible futures of postcolonial citizenship in India. These examples can be multiplied from every continent of the earth and their increasingly connected cities not only through networks of learning machines but also the professions such as architecture, engineering, planning, science and technology entrepreneurs and national, international and transnational authorities traversing both these fields of expertise and fields of application of the learning machines.[7] These learning machines can be designed as cooperative as competitive as different forms of capital and different forms of subjects are accumulated. These competitive struggles over accumulation of capital and accumulation of subjects traverse juridical and legal borders.[8]

All of this is not happening without social and political struggles over the city. The resistances to being incorporated into the learning machines as automated subjects range from protests about technological solutions to urban problems or the negative impacts of these learning machines on human agency, social justice and democratic processes. It is now clear that smart city projects often adopt a reductive, technocratic approach that frames the city as a problem to be solved through data-driven optimisation and control. This can lead to the marginalisation and dispossession of diverse local knowledges, the erosion of political engagement and the reinforcement of existing power structures and inequalities. There are also resistances to privacy and security encroachments, as well as the uneven distribution of the benefits and risks of learning machines.

The critics of the smart city's reliance on cybernetic models of urban management, which reduce citizens to passive data points rather than active participants in shaping their social space, consider whether there can be a more holistic, human-centric approach to urban development that integrates technological innovation with social, cultural and political considerations.[9] This could involve empowering local communities, fostering diverse forms of urban intelligence, and developing counter-logistical strategies to resist the excesses of technocratic control. Yet we are realising that the rise of smart cities is posing complex ethical, political and social challenges, rather than simply creating technological solutions. If indeed an urgent political need is to reimagine the city as a site of democratic citizenship, collective flourishing and environmental sustainability, rather than a machine to be optimised, as many critics argue, we need to understand how this machine is assembled not only from the rationalities of government that mobilise these projects but also from their performative affects and effects that are transforming and integrating the citizen-subject into the autonomic subject of the polity – the autopoietic subject. How are the concepts, practices and methods that arise from and articulated with datasciences and neurosciences assembling the city as a learning machine and incorporating it into systems and networks of transversal polities?

To put it differently, from our perspective it is less important that there is a generally agreed definition of smart cities than what the imaginary of the smart city performs and brings into being as affects and effects of sensory power. The concepts that are hidden in plain sight are often not the ones that dominate the discourse but the ones that function as operating principles to make things happen. We can, of course, discuss the characteristics of smart cities, the role of data and artificial intelligence, the evolution from digital to smart cities and the potential benefits and challenges of building smart cities, and criticise the lack of a universally accepted definition of smart cities, with different perspectives and interpretations. However, we want to ask how things are done with numbers and neurons to build learning machines that function as self-regulating and self-organising polities that produce the autopoietic subject as their affect and effect.

The integration of infrastructure, architecture and technology to design systems such as transport, energy, healthcare and administration are offered as rationalities, but the creation of self-monitoring, analysis and reporting capabilities, as well as the use of data, algorithms and automation to optimise systems as interconnected networks, are carried out by actors, concepts, practices and methods that are performative functions of these systems. If the smart city is seen as a post-Fordist, flexible and adaptable space using digital technologies to coordinate and manage flows of people, goods and information and if this shift is enabled by the proliferation of connected devices, big data and artificial intelligence, we want to ask how cities as species of polities as learning machines are assembled. To enter into a discourse on smart cities, including their rationalities such as sustainability, economic competitiveness and quality of life, is to repeat, iterate and cite their rationalities without fully examining how these rationalities produce performative affects and effects of sensory power. To evade this discourse, we focus on concepts (autopoiesis, sentience, platform, networked), practices (capturing, monitoring, detecting, pulsing) and methods (sensing, identifying, ranking, evaluating) as integrated elements that generate the imaginary of the twenty-first-century city behind its discursive names.

Autopoiesis

We now need to flesh out the concept of 'the autopoietic subject'. Perhaps the key to understanding the imaginary of the twenty-first-century city lies in the concept of autopoiesis, which encompasses the convergence of datasciences and neurosciences. This concept views cities as complex, living organisms that are constantly evolving, maintaining a delicate balance between internal and external factors to govern themselves as autonomous and self-regulating polities. For Christopher Kirwan and Stefan Dobrev (2022) the smart city is autopoietic in this sense. They think that the smart city is a new paradigm of urban systems that leverage the power of AI-enabled technologies to achieve anticipatory, adaptive and self-regulating abilities. They suggest that this new form of collective intelligence, which is a combination of the physical city, human organisational behaviour and machine intelligence, will enable a deeper awareness and symbiosis with other life forms, allowing cities to evolve as autopoietic systems, becoming an integral part of nature itself rather than remaining human-made constructs that deplete nature.

The concept of autopoiesis, introduced by biologists Humberto Maturana and Francisco Varela (1980), describes the essential processes and characteristics that are fundamental for all living organisms.[10] Autopoietic systems consist of self-creating processes that produce all components and subcomponents necessary to sustain their existence as living entities. When this concept is applied to the city, Kirwan and Dobrev say, it can be visualised as complex living organisms composed of multiple functions working together as real-time operating systems. They imagine that the characteristics of autopoietic systems, such as sentience and cognition, structural coupling, ambient intelligence, anticipatory capabilities and self-regulation, can be incorporated into the design, planning and operations of future smart cities.

This new architecture would reflect the dynamic flow of the convergence of spatial and technological patterns, representing the evolution of cities as living, dynamic organisms with the ability to adapt and integrate new technologies within the operational framework and unique culture of each city as a self-generating operating system. Kirwan and Dobrev are convinced that smart city functions, such as smart government, smart people and smart living, can be *designed* to embody the principles of autopoiesis, promoting inclusiveness, enlightenment and actualisation. The platforms providing both public and private services, the pursuit of individual and social enlightenment, and the process of evolutionary convergence will enable humanity to reach a higher dimension of consciousness defined as actualisation will be key components of this new vision for smart cities.

This may be seen by some as utopia or dystopia, but what this concept reveals is the imaginary of the city as a living organism that can be designed. Whatever its rationalities such as enlightenment, actualisation and inclusivity and whatever the heterotopic images, such as anticipatory ability and self-regulation, the design of intelligent agents – which, as we have seen earlier, Turing recognised long ago and Deutsch operationalised – depends on systematically planned and circumscribed conditions under which agents act intelligently, intelligibly and responsibly and the performance of their actions is monitored, measured and evaluated against these goals, which are set independently of their intelligence. These are the essential principles by which sensory power functions. This raises the complex issue for the imaginary of the autopoietic city and its relationship to sensory power. What forces produce this imaginary?[11]

Sentience

The key operational concept in building cities as learning machines is to conceive of them as sentient machines made up of sentient agents. This is the obvious link between datasciences and neurosciences, since the machine and the body are understood as analogous sentient apparatuses. If the central component of a learning machine is a sentient agent, the capacities of that agent before it can be sentient involve the reception and transmission of signals, which are messages for the actuators to interpret and transform into perceptions for the agents. The sending and receiving of signals and their interpretation constitute the perceptual structure of the agents, which is transformed into action structures that actuators trigger.

This is why cryptographic algorithms are the language of authentication and validation of signals to ensure that they come from identified sources and are interpreted without interference.[12] If the physical or social space provides certain options for action, it is the signals that enable those options to be carried out.

Cryptography addresses the challenges of hiding and securing data and information flows, controlling access and exploiting and relying on the properties of communications. If cities as systems of learning machines are characterised by connectivity, interactivity and autonomy, and by the need to extract appropriate knowledge from the vast amounts of data collected by such learning machines in their connective, interactive and autonomous actions, then cryptography becomes the language of cities as learning machines. The city imagined as a system of learning machines not only hears, speaks, sees, touches and tastes; it can also elicit these sensations in other learning machines.

The autopoietic subject becomes confined in the city as a learning machine, both as an agent and as an actuator enabled by cryptographic language. The autopoietic subject is traversed by numerous learning machines: essential elements of social life as apparatuses of government such as pleasure, family, friendship, education, health, travel and work traverse the autonomic subject to constitute learning machines appropriate to their designs. The autopoietic subject is best understood as an agent traversed by these apparatuses and functioning appropriately or intelligently in response to signals triggered and transmitted by actuators. Each autopoietic subject is an accumulation of such apparatuses, and the subject's spatiality and temporality are determined by the specific combination of these apparatuses that traverse the subject. The autopoietic subject becomes something only in a given space and time and is virtually nothing outside its space–time assemblage. The autopoietic subjects are actuated to align the demands and needs of accumulation of different forms of capital with the apparatuses that capture them.

The primary mechanism for achieving this realisation is the platform, which can function as both a verb and a noun. As a verb, platforming subjects means capturing their desires and pleasures as sentient agents and incorporating them into sentient learning machines. As a noun, it describes the specific learning machine that passes through the autopoietic subject: sexuality, family, friendship, education, health, travel and work are often managed, organised and structured through their platforms as sentient machines.

Platform

Can we then describe the city as a learning machine as a platform? This is perhaps the dominant view but it is better to think analytically of the city as an assemblage of platforms or a system of learning machines. Artificial intelligence systems such as autonomous cars, robots and city brains are shaping the social, economic, political and ecological dynamics of cities, but to think of the deployment of digital infrastructures, including sensors, algorithms, databases and mobile platforms, to monitor, regulate and recalibrate the everyday lives of citizens as a platform is probably to give too much power to the idea of the platform as a homogeneous mechanism.[13] These digital platforms and systems, often promoted as increasing efficiency, transparency and citizenship, also control, regulate and normalise people, making them subjects of an attention economy.[14]

The emergence of city brains, large-scale systems that can sense the entire city and make autonomous decisions across multiple domains, such as traffic management, security and planning, exemplified by Alibaba's City Brain, operate through extensive sensor networks and large-scale data processing, navigating the unknown and uncertain real-life scenarios of the city. The spatial dimension of the city brain is rightly described as elusive, ubiquitous yet immaterial, with the ability to transform the city through its actions.[15] But to think of the city brain as a platform is probably an overstatement, since the city can function like a brain only insofar as it assembles multiple platforms that cut across the autopoietic subject.[16] To put it differently, the city is a learning machine, composed of multiple platforms as apparatuses of government and exceeding and transcending the boundaries of the city by connecting it to larger networks.[17]

The multiple platforms that aim to reduce traffic congestion and detect accidents are very different from those that aim to monitor and regulate heat dissipation from buildings.[18] Yet these platforms incorporate citizen-subjects into data-driven decision-making for governing the city with their logics, impacts and effects. As we have seen above, the design of intelligent agents requires systematically planned, goal-oriented platforms as mechanisms for regulating and controlling their agents' behaviour under specific conditions. It is this specificity of intelligent agents that allows the platforms to function appropriately and to enable responsive and rational – autonomous or intelligent – behaviour among the agents when necessary and respond in automated ways when it is required. The imaginary of the city is then not a platform but an assemblage of apparatuses of government increasingly designed as platforms.[19] For cities to function like a brain a conceptual shift is required, which is accomplished by designating the city as an autopoietic system of systems.[20]

Networked

Is the autopoietic city analogous to the artificial neural networks that constitute learning machines? If indeed the growing prominence of networked technologies and digital infrastructures in the design and development of cities reflects an imaginary created by governments, companies and organisations that want to use digital technologies such as sensors, analytics and automation to optimise city functions and services, with the stated goals of improving efficiency, sustainability and economic growth, the city cannot be imagined a closed system of systems but also an open system connected with other systems. William Mitchell (2003) anticipates the smart city as a networked city implemented through large-scale city initiatives around the world. Mitchell is prescient in imagining the city as a networked intelligence but also in describing its effect as cyborgs accepting the influence of such intelligence on people's sense of identity and belonging. Mitchell thinks that events in the digital worlds are increasingly intertwined with physical worlds, and that connections and continuity of people are no longer solely rooted in a specific geographic location but rather in the digital networks that link them to dispersed people and places that are meaningful to them. This shift reflects the growing integration of technology into everyday life and the ways in which it is reshaping individuals' experiences of space, community, and identity in the modern, 'networked' world. While the smart city is described as a symbolic and material construct, with narratives of human betterment through technological innovation, these rationalisations are very different from their realisation through performatives. It seems the imaginary of the city is becoming a combination of open and closed systems that increasingly integrate into a planetary network of systems.

If these large-scale city initiatives are indeed increasingly and in some ways inevitably embedding network technologies into the fabric of cities, mediating city governments, services and everyday life, they are also embedding each city in a network of cities as systems of cities.[21] If this is transforming the sense of place of citizen-subjects, their belonging and civic engagement, with both empowering and constraining implications, then we need to think how the city as a difference machine transforms into a learning machine that operates as a system of open and closed systems, both territorialised and reterritorialised and always connected. If we are witnessing the intensification of datafication and control, as these technologies enable the collection, analysis and automated management of continuous and simultaneous data streams, we also need to understand how sensory power both produces and is produced by the city as a learning machine, and how practices such as capturing, monitoring, detecting and pulsing perform the city as a sentient, autopoietic and networked system of open and closed platforms. While advocating efficiency and optimisation, algorithmic decision-making and surveillance also redefine people's sense of their rights. If the autopoietic citizen-subject of the twenty-first-century city has the anticipatory capacities of self-regulation, self-control and self-organisation, its actual affects and effects seem to be that people increasingly experience the burden of being autopoietic subjects, activated to trigger responses rather than in control of their life designs and desires.

Practices: capturing, monitoring, detecting, pulsing

The practices of instituting the imaginary of the autopoietic city are as diverse as the initiatives themselves. There are autonomous transport systems, urban data management and visualisation technologies, networked infrastructure and services, and the use of emerging technologies such as blockchain, drones and of course intelligent agents. Some of the most debated cases of infrastructure and operations include Masdar City's personal rapid transit system, which uses a network of sensors to guide driverless vehicles along pre-defined routes.[22]

The practices cover diverse domains, including smart mobility, smart healthcare, smart energy, smart community, smart education, smart lighting and smart waste management. The Dublin Traffic Management and Incident Centre and the Padova Smart City project are examples of these smart city applications. The London Data Store, a platform that aggregates and publishes data from the city's operations and transactions to encourage the development of applications and services, is often highlighted as an example of real-time visualisation of diverse city data streams (Kitchin, Lauriault, and McArdle 2018). The potential emerging technologies such as blockchain, digital democracy and smart contracts are also advocated as solutions to problems associated with implementing smart cities. De Castro Neto and Cartaxo (2021) show how blockchain provides a trusted, transparent and immutable record, with applications in areas such as government decision-making and voting systems. The use of drones with remote operators in smart city applications such as traffic monitoring, infrastructure inspection and emergency response is also widely known. Various commercial services use recognition systems with data captured by drones to provide intelligence to planning, development and security agencies. The technologies, practices and challenges involved in the development of smart cities continue to emphasise the importance of data-driven and integrated city systems.

These practices are often justified by practitioners with rationalities such as equality, liveability, sustainability, efficiency, empowerment and engagement; by contrast, critics describe performatives, affects and effects such as control, alienation, estrangement, disempowerment and disengagement. It is necessary to explore these practices for understanding how they are mobilised by sensory power through capturing, monitoring, detecting and pulsing and how these create apparatuses of government in the production of the autopoietic citizen-subject. This requires not only describing these practices in terms of their rationalities and performatives but also relating them to different forms of power, especially sensory power.

Capturing

It is by now well understood that the use of digital technologies and machine learning in cities, particularly in the imaginary of smart cities, requires massive data acquisition and monitoring infrastructures and sociotechnical arrangements (Aydın 2022, 16; Sullivan 2022). The sensors and data-gathering systems are used to capture various types of data, such as population density, environmental conditions and human activities (Alavi et al. 2022). This enables the datafication of movements, flows and states of cities.

De Castro Neto and Cartaxo (2021) describe how cities are increasingly being captured through various technological means. Flows of resources, products, people and information throughout urban infrastructures are being digitised and datafied through sensors embedded in ICTs, transport, water, energy and waste networks. The states of cities as physical and social spaces, such as population density, air temperature and quality, light and sound levels, are being monitored through sensors, cameras and synoptic instruments like satellites. Meanwhile, connected machines and devices used for personal and professional activities in cities allow the measurement of transactions, consumption and communication patterns, including people's activities, communication and interactions, the interactions between people and their cities, and the interactions among components of their physical and social space such as communicating and autonomous machines and devices.

Yet the practices of capturing are also about producing autopoietic citizen-subjects. When we say production we do not mean only their symbolic but also their material realisation: autopoietic citizen-subjects are increasingly becoming integrated bodies assembling various sensory apparatuses as intelligent agents. Gabrys (2019; 2014) provides an example of how citizenship in smart cities is transformed as smart citizenship and citizens as citizen scientists. The programming of physical and social spaces through sensor technologies has transformed conceptions of the city and citizen participation. The sensing citizen is presented as the ideal mode of citizen engagement in smart cities, where citizens are required to participate in the flow of digital information. If this raises questions about urban exclusion and the extent to which citizen sensing leads to meaningful actionable policies, as the monitoring technologies may co-opt citizens into reactive modes of self-government, it also shows the importance of how power is exercised differently. Gabrys suggests that the programming of cities through sensor networks and algorithms generates new political techniques for governing everyday life, where citizen engagements and urban processes unfold through these spatial and temporal networks.

The capturing of subjects is about producing them as autopoietic subjects. Ayona Datta shows the relationship between the production of smart citizens as autopoietic subjects in postcolonial cities, which required technocratic systems envisioning, planning and restructuring almost 100 cities (Datta 2018). The accumulation of subjects and the accumulation of capital through sensory power requires special apparatuses of capture. These apparatuses such as education, finance, health and work deepen their reach through sensory power to produce the autopoietic subject and city as a learning machine.

We can divide the apparatuses of capture that make particular use of sensory power into two broad processes: datafication and subjectivation. The process of datafication requires the acquisition and monitoring of data through sensors but it must be linked to forms of capital. The acquisition is often driven by the will to power for symbolic, cultural or social capital that can be converted into economic capital. This is when the rationalities for data acquisition are transformed into imperatives to capture data first and reason later. To achieve this, the accumulation of subjects is necessary: the design of intelligent agents includes the inclusion and embedding of humans as intelligent agents in these systems. The discourse on the digital divide as inequalities in access to digital technologies often obscures the double process of accumulation of capital and accumulation of subjects and of datafication and subjectivation.

The apparatuses of capture that perform for disciplinary and regulative power now have access to considerably more detailed data and surveyable subjects with constant monitoring. More importantly, apparatuses of capture also have now access to both movements (location) and activities (action) of its subjects continuously, simultaneously and instantaneously. There are numerous examples of how sovereign, disciplinary and regulative forms of power are being augmented by sensory power with its access to location and action. The apparatus of education increasingly uses machine learning techniques such as adaptive learning, virtual reality, learning style analysis and academic performance prediction. With the promise of a learning process that is more effective, personalised and accurate, the apparatus of education also tracks, traces, monitors, evaluates and credits learners through location and action data. The workers are also being tracked and traced for both their productivity and their creativity, augmented by location and action data deployed by sensory power.

The apparatuses of capture for both accumulation of subject peoples and accumulation of forms of capital have never had as much access to subjects and objects of power as now they have been augmented by sensory power. The practice of monitoring and detecting of all the elements of a given apparatus continuously, instantaneously and simultaneously enable learning machines to become pulsing systems not like but as nervous systems. This is not by accident but by design.

Monitoring

The apparatuses of capture for disciplinary power and regulative power always depend on ongoing monitoring such as surveying. These apparatuses are augmented by the instantaneous monitoring that sensory power enables. The most visible and impressive manifestation of apparatuses of capture for sensory power is the emergence of dashboards and control rooms akin to airports or space centres. We are indebted to Kitchin, Lauriault and McArdle (2015a, 2015b), Kitchin and McArdle (2018) and Kitchin, Coletta and McArdle (2020) for providing us with images of how city dashboards employ visual analytics, such as dynamic and interactive graphics, infographics, maps and multidimensional models, to display and communicate information about the performance, structure, patterns and trends of cities. These dashboards increasingly incorporate real-time data from sensors, actuators and crowdsourcing, in addition to traditional sampled data. Although the scope of data displayed in dashboards is limited, and there are concerns about the trustworthiness and validity of the information they present, what they reveal is the desire for instantaneous, continuous, live monitoring.

Such monitoring systems are now considered vital for structural health monitoring and early warning in cities, relying on the structural indicators of the buildings themselves. Water quality monitoring, for example, requires an architecture for the development, deployment and commissioning of a multi-node water quality monitoring system using complex protocols to regulate the quality of drinking water and traversing national and international regulations.

The integration of intelligent agents in cities are accomplished by control rooms that can generate, process, analyse and act on instantaneous data. These control rooms are shifting from human-in-the-loop configurations to more automated forms that utilise computation to monitor and regulate systems in automated, automatic and autonomous ways, with decision-making delegated to intelligent agents. The effect of these monitoring systems is the design of infrastructures for integrated with information and communications technology for distributed structural sensing systems that serve as autonomous intelligent agents to measure various physical or social space parameters, react in time, establish automated control and collect information for intelligent decision-making.

Detecting

If indeed the control rooms are shifting from human-in-the-loop configurations to more automated forms but including humans and non-humans in operations, this requires the performances of intelligent agents as actors. Various technologies, including the connected devices and sensors, actuators and agents for monitoring and detecting infrastructures, movements and actions across cities are constantly creating programmable actions and protocols that transform decisions from human-in-the-loop into human-as-the-machine configurations. This is clearly decisive in the birth of the autopoietic subject that is neither automatic nor autonomous but autonomic.

The instantaneous detection of traffic accidents, the use of programmable and multifunctional materials for structural integrity monitoring, data fusion and pattern recognition techniques for structural health monitoring, the potential to add evaluation and judgement by annotations to gathered data through machine learning or crowdsourcing, personality recognition, the use of mobile agents for mobile crowdsensing campaigns, the use of intelligent machines for predictive analysis and decision-making in planning, the use of object recognition algorithms to recognise different elements of city infrastructures and superstructures, and the coordination of control devices such as drones through mesh-based networks and cloud-based intelligence are examples of detection and repair technologies that work with humans as intelligent machines among non-human intelligent agents, each performing their autopoietic functions.[23] Taken together, this arguably institutes the city as a pulsating living organism, the city as a learning machine.[24]

Pulsing

The affects and effects of capturing, monitoring and detecting practices increasingly transform the city as a pulsating (living organism) system of learning machines.[25] This is the equivalent of calibrating the body with population through regulative power. For sensory power, however, the question is not about calibrating bodies but incorporating them as intelligent machines into clusters. The performance of bodies is adjusted to specific clusters assembled to achieve specific rather than a population. The autopoiesis of the city and its machinic agents constitute it as a living organism just as the imitation game imagined. If Logani and Makkar (2024) formalise this imaginary of the city by outlining machine learning algorithms as the structure of the city and Dimitrios Amaxilatis and his colleagues (2020) articulate this as the learning city, we can see how the city is imagined already as a performable if not programmable autopoietic system, an engineered neural system composed of numerous clusters traversing bodies. Each autopoietic subject is composed of a unique, dynamic and fluid combination of clusters as bodies continuously fulfil their obligations and exercise their rights in becoming subjects of polities as learning machines.

The importance of pulsing as the sending, reception, transmission and processing of signals in the presence of others and coding and decoding them as a living being in the presence not only of other human but also non-human intelligence is the operative principle of polities as learning machines and the autopoietic subject. The autopoietic city integrates and embodies human and non-human agents, merging them into a unique form of intelligence. The nature of this intelligence, whether mechanistic or autopoietic, and the specific sense in which it manifests, remain open questions. What affects and effects sensory power produces as pulsing machines is perhaps encoded in cryptography, not as a metaphor or allegory but as a technology that renders pulsing machines intelligible and sensible. When cities as system of intelligent machines are advocated as a frame for intervention around concerns of optimisation, rationalisation and coordination, capitalising on city impulses while tamping out excesses, the pulsing machine appears certainly and object government. With the increasing use of materials like piezoelectric transducers and carbon materials as sensing materials or other self-sensing composites with technologies for real-time semantic annotation of streaming data, we increasingly see the imaginary of the city as a learning machine produced by sensory power.

Methods: sensing, identifying, ranking, evaluating

The most important point we can make at this stage before discussing how methods produce the autopoietic subject is this: methods performed to realise concepts and mobilise practices in the production of the imaginary of intelligent cities as systems of learning machines arise from and make use of methods invented by the datasciences and neurosciences. The accumulation of methods, from classification and clustering in the datasciences to diagnosis and recognition in the neurosciences, is the backbone of the imaginary of polities as learning machines, of which the city is a special case. The various methods and models of statistics and analysis in the datasciences and parallelism and plasticity in the neuroscience are prevalent in the city as an intelligent system of agents. This is not to say that the challenges of producing the imaginary of the city as a learning machine have not imposed new necessities on the invention of sensory power. On the contrary, this field of application has invented new methods by combining previous ones. For example, blockchain technology and the application of ant colony systems have generated answers to novel problems of the city just as vehicle routing problems or rapid authentication problems generated sui generis methods. Similarly, the invention of potential attraction and repulsion fields in multi-agent systems, the importance of feature selection and engineering in the selection and collection of data for analysis are unique problems arising from the complexity of the city as a living organism or autopoietic system.

Perhaps most outstanding or overwhelming development has been the proliferation of open-source methods that combine datasciences and neurosciences. The Hadoop ecosystem, including Hadoop Distributed Files System, MapReduce, Spark, Storm, Apache Flink, H2O and Kafka, is an example of online (open, semi-open and commercial) methods that provide distributed storage, parallel processing, stream processing and machine learning capabilities for handling large-scale and real-time data. These methods are available to almost anyone who has access to large-scale data and technical knowledge. The ways in which they verify, prepare and analyse large-scale data are staggeringly complex. But it is possible to develop a typology of these methods for their performative functions in producing cities as learning machines: sensing, identifying, ranking and evaluating. These are the methods that enable affective and effective strategies and technologies of sensory power and are available to those who have mastered its knowledges.

Sensing

As we have seen, the development of cities as systems of intelligent agents in the twenty-first century is enabled by sensory technologies, often networked, to regulate and automate urban processes and resources such as transport, buildings, electricity and industry, with rationalisations of improving the efficiency and sustainability of the system. Often the sensor data are collected, aggregated, visualised and used for instantaneous and continuous analytics and prototyping. We have also seen that data capture is done by special purpose sensors as well as general-purpose sensors found in devices most people use to negotiate their lives in the city. This latter form of sensing has given rise to a type of sensing named as crowdsourced sensing. It interpellates citizen-subjects to act as sensors by contributing data through their devices with or without their knowledge or awareness. This enables distributed and collaborative data collection, processing and knowledge creation about a specific object of sensory power. It is tempting to interpret the city as a learning machine composed of all sorts of robots but this should be resisted as long as we understand robots as human or animal sentient beings. Instead, as we have seen, intelligent agents include a range of technical actuators, switches, software and hardware. As Studley and Little (2021, 72) observe, the city as a learning machine 'senses, processes, and acts, becoming an embodied intelligence evolving in symbioses with the other actors in its ecosystem'.

Crowdsourced sensing has led applications in navigation and movement, social and cultural life, public services such transportation, health, welfare, education, learning, culture and sports (Leppänen et al. 2017). If we are now used to aerial sensing and monitoring systems, such as drones, inspecting infrastructure operations for more cost-effective condition assessment, we are also conditioned to see them as benefits. We participate in processes that simplify extracting knowledge and annotations from urban data streams, leveraging machine learning and crowdsourcing, and make crowdsensing increasingly affective and effective. These crowdsensing technologies, data analytics and human–machine collaboration are transforming cities into 'embodied intelligences' that can sense, process and act on impulses to support decision-making and regulate services and infrastructure. All this is not only mobilised by sensory power but is also produces the object – the crowd – that it ostensibly senses. But the crowd of sensory power is fundamentally different from the crowds of disciplinary or regulative power: its affective and effective principle is not to measure and adjust bodies to norms (disciplinary) or metrics (regulative) but to generate functioning clusters as learning machines for specific purposes. The sense of a crowd in disciplinary power affects a collective formation (protests, gatherings, assemblies); in regulative power it affects a movement (petitions, objections, collaborations); and in sensory power it affects a cluster (nameless, fluid, dynamic).

Identifying

When networked sensor technologies become embedded in the city, identifying objects, agents, movements and actions in cities becomes essential for sensing the city for enacting protocols and procedures. The identification methods for sensing the city can be divided into primary and secondary methods.

The primary methods of intelligent identification agents include recognition and imaging. Take, for example, the identification of movement: of concern is the objects or subjects that are moving, and their direction, speed and pace of movement. This often requires automated airborne sensing if the moving objects and subjects are not equipped with sensors, otherwise sensors such as accelerometers, location identifiers, gyroscopes, proximity sensors and magnetometers are used. Yet identifying movement does not accomplish the operation of identifying the actions of both objects and agents. This gives rise to secondary intelligent identification agents for labelling or annotating the movement. It is this labelling and annotating that converts quantities recognised by intelligent agents into qualities. This is where secondary methods of identification come into play as intelligent agents train models without knowing the operative principles determined by designers. These methods use biometrics and neurometrics (Ambika, Radhika, and Seshachalam 2020).

The objects, agents, movements and actions are brought together and are identification methods are performed on the results of the primary identification with secondary identification by not only labelling and annotation but also inference and judgement. Labelling or annotation methods are complex and depend on the available categories and classes of movement or action. In crowdsourcing methods, some agents themselves are involved in identifying their own actions by responding to impulses generated by sensing systems. The inference methods are even more complex, as the movements or actions are interpreted using machine learning methods (Alavi et al. 2022).[26]

Ranking

The methods of ranking are crucial for transforming quantitative data into qualitative knowledge. All elements of sensing (both direct and indirect) and identifying (both primary and secondary) objects, agents, movements and actions can be ranked against design objectives. The systems of intelligent machines composed of these elements can be ranked against performance indicators. Cities as intelligent systems of learning machines can also be ranked against performance measures with data captured from all other cities in the system.

These methods are already operative. The European Union has officially recognised the concept of 'smart cities', with the European Parliament conducting a study that ranks cities based on their performance such as government, affordances, liveability, mobility, economy and environment (De Castro Neto and De Melo Cartaxo 2021, 58). The underlying assumption is that the imaginary of smart cities is rooted in the creation and connection of different forms of capital (social, cultural, symbolic, economic) and information and communication technology infrastructure, with the stated aim of 'generating greater and more sustainable economic development and a better quality of life for citizens' (Studley and Little 2021, 75). Such methods of ranking are mechanisms of creating systems of cities as systems of learning machines.

Evaluating

The methods of evaluation are multidimensional ways of ranking to transform quantitative data into qualitative data that can be acted on for decision and judgement. The use of machine learning algorithms for credit scoring and the integration of user interaction with the OrganiCity platform deploys such methods of evaluation. The various machine learning algorithms, including XGBoost Classifier, Bagging Classifier and Support Vector Machine, on an Australian dataset of loan applications or the Chinese citizenship credit scoring evaluate intelligent agents from the point of view of systemic performance. Similarly, the interaction between users and the JAMAiCA system, which is facilitated by OrganiCity tools like the Urban Data Observatory and the Sensing on the Go smartphone application, allow users to discover and filter relevant assets based on factors such as geolocation, types, metadata and reputation or recommendation scores.

The concepts (sentience, platform, autopoiesis, network), practices (capturing, monitoring, detecting, pulsing) and methods (sensing, identifying, ranking, evaluating) are operative together as an apparatus that produces the autopoietic subject.

Governing polities as learning machines

How datasciences evolved from statistics, analytics and machine learning, inventing new concepts, methods and practices, articulating new objects and, of course, producing new subjects, how the emergence of analytics from statistics inaugurated the population game and regulative power, and how the emergence of computation, modelling and learning inaugurated the imitation game is an intricate story. The population game and the imitation game are not identical games; it is their differences that play out forms of power – especially regulative and sensory – but they are nonetheless internally related in multiple ways. First, sensory power would not have been possible without regulative power; the latter depends on the concepts, methods and practices developed by the former. Second, both forms of power now derive their effectiveness from being exercised simultaneously. Their logics straddle each other. Third, both their subjects and objects are interdependent. If the population game involved calibrating between populations and bodies through regulative power, the imitation game involves learning machines as autopoietic polities.

We summed up the convergence between datasciences and neurosciences in one pithy phrase: from machine learning in datasciences to learning machines in neurosciences. This development has been anticipated in many ways since the 1930s, but its results are experienced and felt most strongly in the twenty-first century. Now machine learning incorporated techniques such as probability and stochastics, developed clustering methods with supervised, unsupervised and partially supervised algorithms, and produced new statistical methods such as cluster analysis and correspondence analysis. The originality of these methods is not only in the specific techniques, which were most ingeniously developed, but also in the way they gradually shifted the purpose of analysis: from explanation and prediction to diagnosis and autopoiesis. The application of these technologies of power to massive and dynamic stores of data assembled from networked devices, especially in the last twenty years, has made machine learning possible.

How machine learning in datasciences precipitated the development of learning machines in neurosciences is immensely complex. But, as the development of computational neuroscience illustrates, the application of machine learning techniques became remarkably adaptable for the production of intelligent machines as learning machines. The design and production of intelligent agents as learning machines largely relies on machine learning methods such as cluster and correspondence analysis, modelling and simulation.

A new subject and object of government emerges from this transformation. The autopoietic polity is a learning machine that functions like the brain, at least as the human brain as now imagined by the field of knowledge that drives sensory power: computational neuroscience. To repeat, we use the polity (or polities) as it was used in European political and legal discourse since the twelfth century: a multitude of people combining and integrating into a collective body with a generative will greater than the sum of the constitutive wills of its parts. At the original scene of ancient and modern polities were political agents as autonomous subjects. At the original scene of learning machines are intelligent agents as autopoietic agents. Doing things with sensory power today means designing and building autopoietic agents, equipping them with sensory apparatuses appropriate to their domains of operation, creating networks integrating these agents, and enabling these networks to perform generative and reparative functions. Controlling these learning machines begins with their design, which means building them for specific performative functions operative in specific domains. If polities were originally multitudes such as cities, states and empires, polities as learning machines are connectomes with transversal territories, multitudes and networks. The polities as learning machines are deterritorialised and traverse territorialised polities such as cities, states and empires but also reterritorialise them.

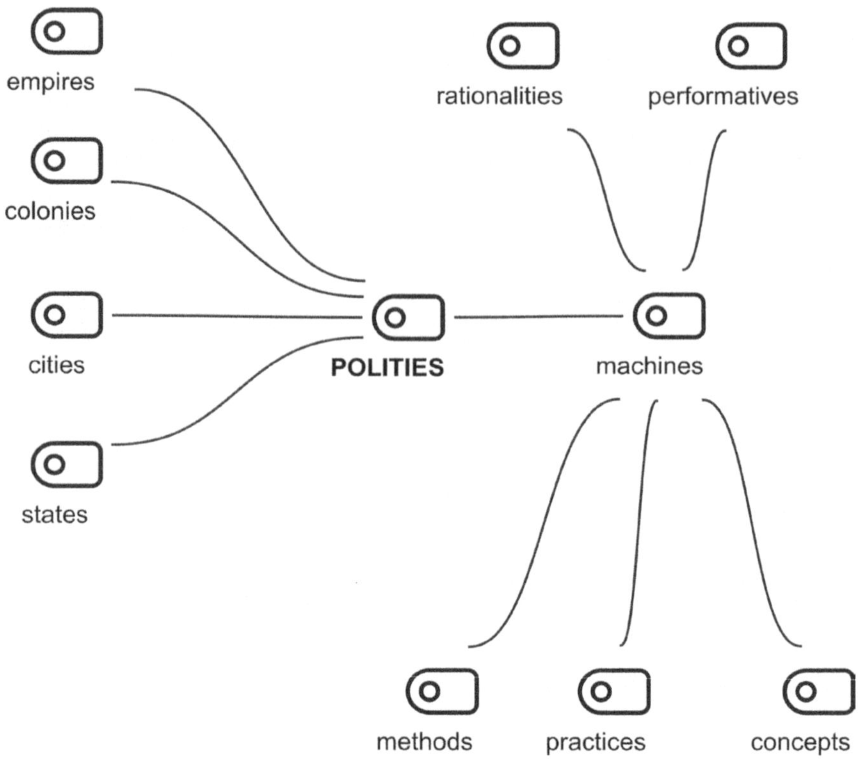

Figure 4.2 Polities, cities, states, empires and machines

There are multiple ways of responding to this perplexing question that we gradually articulated. We respond to it by examining perhaps most powerful enactment of sensory power: the city. All forms of power have generated an imaginary of the city appropriate to its affects and effects, strategies and technologies. The city has been a primary assemblage of power for some 6,000 years and other polities such as states, colonies and empires have always built upon them. Lewis Mumford (1938; 1961) is known for his most evocative and provocative interpretations of the city, but his later and lesser-known works on the city and the machine outline a modified relationship between power and the city ([1966] 1967; [1964] 1970). What Mumford has achieved in these later works that his focus moves away from the dominant narrative of the city as an agglomeration or settlement to the city as a politically instituted imaginary. When he describes the modern megalopolis as the new megamachine where most functions of the city were being integrated into cybernetic systems, he sees the city in history in a way that his earlier work only probes. The new image of the city is a deterritorialising polity traversing territorialising polities such as empires and nations but also reterritorialising them.

The birth of sensory power has already generated a powerful imaginary of the city, with competing concepts such as the computable city, the cryptographic city, the cybercity, the digital city, the information city, the intelligent city, the networked city, the platform city, the sentient city, the smart city and the wired city. These can be seen as variations on the megamachine with a twenty-first-century iteration – an iteration that would have been unthinkable without the convergence of datasciences and neurosciences. To appreciate this twenty-first-century imaginary, an interpretation of the city as a learning machine will be necessary as it is embodied in the inventions of the citizenship game.

Notes

1. There are several fields of scholarship that deconstruct the Eurocentric or orientalist narratives: the origins of the state, the origins of government, human migrations and world history (Manning and Trimmer 2020; Yoffee 2015; Clark 2013; Lucassen, Lucassen, and Manning 2010).
2. The title of the French edition of his book, *Homo Domesticus* ([2017] 2019), captures the essence of his argument that the invention of the polity involved the invention of domination of humans over human and other species.
3. See Scott (2019) for his response to some of his critics.
4. By interpreting the city as a difference machine we are drawing on *Being Political: Genealogies of Citizenship* (Isin 2002). But there are differences. First, we are exploring the polity as a difference machine and seeing cities, states and empires as specific territorialisation, derritorialisation and rerritorialisation of the polity as an apparatus of government. Second, we are also reinterpreting the polity as a difference machine with a reworked conception of power (in) theory and power (in) practice. These two differences extend beyond drawing from *Being Political* but amount to reworking its argument. For more details of this reworking of the citizenship game see Isin (2024).
5. The discourse on smart cities clearly advances the former scenario, where a critical scholarship raises fundamental questions concerning smart cities. There are many examples but the following are typical for their positive outlook: Amaxilatis et al. (2020), Batty et al. (2012), Doherty (2023), Goyal et al. (2024) and Shayan and Kim (2022).

6. It is agreed by everyone that robots and autonomous cars are becoming an integral part of cities, taking on roles traditionally performed by humans. These urban artificial intelligences are both visible, with their physical presence in the built environment, and relatable, owing to their familiar shapes and functions. However, more elusive forms of urban intelligence, like city brains, are also emerging. City brains are advanced as systems that can sense and make decisions about an entire city, processing vast amounts of data from sources like surveillance cameras. Unlike autonomous cars and robots, city brains operate at a much larger scale, with an immaterial, ubiquitous presence across the urban landscape. While automation is confined to controlled environments, the autonomy of city brains requires them to make decisions in the face of persistent uncertainty and unexpected situations. The spatial dimension of city brains is complex, as they are simultaneously everywhere and nowhere, with key physical components like cameras and traffic lights distributed throughout the city. For Federico Cugurullo (2021), ultimately, these various forms of urban artificial intelligence have the capacity to transform the environments in which they operate, with autonomous cars and robots acting as extensions of a single intelligent machine controlling multiple bodies. Most critical literature on smart cities overlook the development of different forms of power (sovereign, disciplinary, regulative) and an articulation of a new form (sensory) that augments them (Halegoua 2019; Mattern 2021; Mitchell 2003; Willis 2020; Shapiro 2020).
7. For critical and interpretive studies on these developments see Coelho et al. (2019), Aydın (2022), Cunha (2021), Goyal et al. (2024), Landahl (2021), Lepakshi and Ranga (2024) and Oliveira et al. (2019).
8. Fleur Johns (2017; 2021; Johns and Compton 2022) has been most useful for making this point.
9. Myria Georgiou (2024), for example, argues for a human-centric approach to digital cities without anthropocentrism.
10. We do not have space to discuss two other trajectories of the concept, the work of Gunther Teubner (1988; 2019) and Niklas Luhmann ([1997] 2012; [1997] 2013), but we are indebted to them for our concept of the autopoietic subject.
11. Kitchin, Coletta and McArdle (2020) recognise that smart city technologies, which include city operating systems, integrated control rooms, intelligent transport systems and sensor networks, have become increasingly prevalent in the twenty-first century. But they warn that these technologies enact social control, as they function through dense and

simultaneous feedback that modulates the performance of infrastructure and those captured within it. For Kitchin, Coletta and McArdle, the embedding of computation into the fabric of cities and its use to manage city services and infrastructures are shifting governmentality from disciplinary forms to those of social control. This raises questions about how these technologies and initiatives, and their associated governmentality and logics of control, fit together and whether they work in concert or in opposition to one another. But it is equally important to avoid a flattened conception of power and assume that sensory power simply reproduces governmentality of disciplinary and indeed regulative power.

12. Richard Coyne (2023) shows how cryptography plays a crucial role in shaping urban infrastructure and communication. He defines cryptography as the art of securing information through encryption but not as a new phenomenon in cities, as communication has always involved hidden messages and coded exchanges. Rather, cryptography is a major component in the perennial life of any city, as it applies to securing transactions, protecting data flows and enabling secure communication among people and digital devices. However, cryptography goes beyond just obscuring plain text messages; it also conceals, protects and secures any data, process, protocol or method of access in the digital domain. For Coyne the smart city, where sensors, actuators and algorithms, are used, optimising the city and enhancing the experience of citizens would not be possible without the secure financial transactions, data flows and communications performed by encryption. The challenge of cryptography in the city is less about coding and hiding and more about decoding and exposing the hidden aspects of urban life and form.
13. Myria Georgiou (2024) accepts that platforms and digital infrastructures are transforming urban societies, altering how people work, socialise and imagine cities. But she adds that these technologies are situated within wider economic, social and governmental systems that order cities and thus technocentric conceptions that privilege platforms as uniquely powerful processes overlook the long histories of urban struggles related to class, race, gender and space. Rather than viewing platforms and digital infrastructures as inherently progressive or threatening, they should be understood as embedded within complex sociotechnical systems that shape urban life in multifaceted ways and that we must be attentive to the nuanced and contextual impacts of these technologies on diverse urban communities.
14. Although Yves Citton (2016) does not mean it this way, his description of the attention economy has strong resonance with the autopoietic subject of learning machines.

15. Jianfeng Zhang et al. (2019) provide a description of Alibaba's City Brain project, which is an end-to-end system designed to extract meaningful insights from the vast amounts of heterogeneous data generated in cities. The system deploys machine learning technologies and computing power to govern and manage the living in cities. City brains have been created in several Chinese cities. Their capabilities across the spectrum from cognition to optimisation, decision-making, search, prediction and intervention constitute the backbone of the system. The overarching goal of city brains is said to improve the way cities are managed and experienced by their inhabitants, harnessing the power of big data and advanced technologies. See Cugurullo (2021) and Karvonen, Cugurullo and Caprotti (2019b) for critical analyses.
16. For example, when analysing city brains, Cugurullo (2021) argues that they exhibit autonomy because they can make decisions independently in the face of uncertainty and unexpected situations, unlike automation, which is limited to controlled environments. This raises the question of whether city brains are autonomous in the way that Cugurullo describes, or automated, as some argue. The spatial dimension of city brains is elusive, as they are ubiquitous yet immaterial, existing as a learning machine rather than a physical entity alone. It is best to resist the temptation to interpret the city brain as an autonomous or automated machine.
17. Mattia Frapporti (2024) provides a connection between platforms as infrastructure and platforms as a form of power for accumulating subject peoples.
18. Hudson-Smith, Hügel and Roumpani (2020) discuss, for example, the street as a platform, which refers to the street being immersed in a dynamic, data-rich environment. This includes not only well-established sources of data such as electromagnetic radiation, radio waves and police voice traffic but also a twitching, pulsing cloud of additional data that characterises the city. They suggest that the street itself can be viewed as a platform, a hub of diverse data streams and information flows that shape the urban landscape and experience. This image is certainly a constitutive element of the imaginary of the city as a learning machine and we will see later that indeed the city is increasingly functioning as an integrated system of open and closed systems assembled from a multitude of platforms.
19. Mattern (2021) suggests that digital platforms can facilitate citizen participation in municipal governance or enable algorithmic decision-making to replace the complexities of democratic processes. However, the limitations of data-driven planning approaches should be acknowledged, and these computational methods should be combined with other ways of under-

standing and mapping cities, such as viewing them as bodies, machines, or ecological systems. Mattern says that this integrated approach is more likely to yield effective responses to pressing urban challenges.

20. Shapiro (2020) shows that predictive policing platforms that use machine learning algorithms with historical crime data and other information to predict where and when crimes will occur is often rationalised or justified with the aim of allocating police resources more efficiently. However, Shapiro rightly cautions that these platforms often fall short of the promised benefits and can have negative consequences for communities. If the imaginary of the smart city fetishises precision, control and logistical efficiency, the frictions, uncertainties and cultural politics inherent in technological interventions reveal the limitations of these ambitious visions.
21. See Castells (1989; [1996] 2011) for an early interpretation of the city as an information network.
22. See Kitchin, Lauriault and McArdle (2015a; 2015b) and Kitchin, Coletta and McArdle (2020).
23. Abraham et al. (2024); Alavi et al. (2022). For mobile agents for crowdsensing campaigns, executing tasks autonomously, leveraging device resources and addressing data collection/participant issues and simulation and real-world evaluation, see Leppänen et al. (2017).
24. See the learning city concept advanced by Dimitrios Amaxilatis and his colleagues where the city is imagined through its performability – a more advanced concept than programmability (Amaxilatis et al. 2020).
25. By adopting some of the terminology of semiotics Richard Coyne (2023) says that 'a message is a species of signal. Signals are messages that are yet to be interpreted by human or other sentient agents. In the digital domain, algorithms check, authenticate, certify, and validate digital signals by means that often involve encryption. Cryptography applies not only to obscuring plain text messages, but also to concealing, protecting, and securing any data, process, protocol, or method of access in such a way as to render it opaque to a majority of human, animal, and machine agents.'
26. Vitaliy Pavlenko and Tetiana Shamanina (2023) outline possibilities of making inferences on emotional states from identification methods. They describe the development of methods in human neuropsychological research. These methods use eye-tracking technology and computational processing to monitor and identify the state of cognitive processes. The aim is the construction of classifiers of human cognitive states, which are built on the basis of heuristic features that are resistant to computational errors.

Conclusion: what games shall we play?

If I'm design'd yon lordling's slave
By Nature's law design'd,
Why was an independent wish
E'er planted in my mind?
If not, why am I subject to
His cruelty, or scorn?
Or why has Man the will and pow'r
To make his fellow mourn?
Man was made to Mourn (1786) – Robert Burns (1993, 54)

The central question that arises from this book is less daunting than Burns's question: what games should we play in the face of the emergence of a new form of power in the twenty-first century? We will need to grasp how various movements, such as datapolitics and neuropolitics, confront concepts, practices and methods that institute an imaginary of the polity as a learning machine.[1]

By tracing the birth of sensory power against a long history of power, we try a strategy that builds on datapolitics and neuropolitics but reworks and stretches them. We organise this long history around games of power we play, which are the kingship, citizenship, population and imitation games, whose births can be roughly dated to 6,000, 2,500, 500, 250 and 100 years ago. All these games are at play today but their playbooks have dramatically changed with the imitation game. Doing things with sensory power today entangles sovereign, disciplinary and regulative forms of power but in ways that make possible to invent new games.

We explore these games with Austin's speech acts. Austin opposes the propositional logic used in machine languages (if, then, else) with performative logic used in life (in, of, by). Austin was focused on doing things with words. We stretch Austin for theorising doing things with numbers and doing things with neurons, which are ways of doing things with power. This certainly requires reworking Spinoza, Nietzsche, Deleuze and Foucault on power by distinguishing between power (in) theory and power (in) practice. While we are familiar with doing things with words, speaking of doing things with numbers and neurons may sound strange.[2] We make them more familiar by speaking of a new form of power – sensory power – when it became most sensible and intelligible during the coronavirus pandemic.

Observing some elements of sensory power during the coronavirus pandemic reveals a strange yet familiar language of clusters, dashboards, tracking and tracing, and it becomes obvious that a convergence between datasciences and neurosciences as two forms of knowledge is at work. How do we understand the history of this convergence?

What the coronavirus pandemic makes sensible and intelligible is the birth of sensory power. The language of hotspots, bubbles and clusters, how these clusters redraw boundaries and borders because they are transversal, how they were connected, how they were visualised, and how they were monitored by their pulses and signals provide the openings of an exploration of the birth of sensory power. We retrace the rise of datasciences from the seventeenth to nineteenth centuries, neurosciences from the nineteenth to twentieth centuries, and their convergence in the twentieth and twenty-first centuries, creating novel types of polity and subject.

The population game precipitates datasciences, which is the primary way of doing things or exercising power with numbers. It emerges in two stages. First, from the seventeenth to eighteenth centuries, disciplinary power subjects and punishes bodies according to the needs of population, wealth and territory of states, empires and colonies. This makes sovereign power somewhat benevolent and anticipates regulative power. Second, from the eighteenth to the nineteenth centuries, population begins to acquire vital signs such as fertility, mortality, longevity and reproductivity, which gives rise to demographics and statistics. The datasciences calibrate bodies and population for mutual adjustment and create a new logic of the whole (population) and its parts (bodies). This is the logic of regulative power. The operative concepts of sovereign and disciplinary forms of power – surveillance and classification – become transformed with datasciences as monitoring and sorting through regulative power. The new concepts of correlation and correspondence, however, from the nineteenth to the twenty-first centuries signal a drive to discover underlying principles of any group (cluster) that bodies will form. With practices of surveying and censusing, sovereign, disciplinary and regulative forms of power undergo a transformative shift, with a renewed focus on inquiry and data mining as means of governing bodies as groups (clusters). This is the incipient logic of sensory power.

The practices of especially regulative and sensory forms of power were made possible and generated successive methods such as statistics, analytics, clustering and learning. The underlying logic of each method undergoes a significant transformation. Statistics, which originally involved fitting samples into populations (calibrating bodies and populations), increasingly shifts towards understanding the underlying and dynamic qualities of samples (clusters) without testing hypotheses about populations. Clustering, in turn, develops specialised stochastic and heuristic modelling techniques that are now widely used in machine learning.

A few scholars traced these transformations but not how datasciences and neurosciences converged. For the birth of sensory power, the decisive moment is the shift in datasciences from probabilistic to stochastic and then to heuristic analyses of the whole and its parts and how this has organised neurosciences. This is the birth of the imitation game.

The imitation game, named by Alan Turing, was invented in the 1930s. It is a pivotal game that crosses over from datasciences to neurosciences. If datasciences had advanced to the point where probabilistic logic led to stochastic and then heuristic logics, enabling the development of learning models, the question of whether machines can think would have arisen. This question profoundly impacts both our understanding of the nervous system, particularly the brain, and our perception of machines, transforming them from mere calculators to thinking machines. The imitation game signals the convergence of datasciences and neurosciences, complicating the question of whether the brain is modelled after the machine or whether the machine is modelled after the brain. Or is it the conjunction of both?

This conjunction is most evident in four concepts that are shared between the brain and the machine: parallelism, modularity, connectivity and plasticity. It is now affirmed that the brain processes information not sequentially but in parallel, communicates with various autonomous or semi-autonomous clusters of neurons operating as connected modules, and is malleable or programmable. We affirm that computational neuroscience and computer science are not coincidentally sharing the same language. If there were a singularity between the brain and the machine, this was not predicting some future event but was a design in the origins of sensory power.

These concepts organise three effective practices shared between datasciences and neurosciences. The diagnostic practices identify any potential risks, dangers, malfunctions or anomalies that require attention in the functioning of any machine–brain. The recognition practices integrate artificial and natural sensory apparatuses. And the automation practices are designed to correspond the results of diagnosis and recognition practices, and either recommend paths for action or perform such actions.

The methods of datasciences converge with neurosciences and generate computation, modelling, visualisation and machine learning methods. Computation focuses on manufacturing and programming techniques to both increase the capacities and improve the design of integrated systems as intelligent agents. The simulation methods enable convergence between machines and brains and visualisation becomes a method of diagnostic and inference. The claim is that machine learning models not only imitate the brain but also aim to create hybrid brain–machine systems to learn and make decisions as intelligent agents.

If we reverse 'machine learning' to name the invention of the learning machine, this is because we see it as an essential outcome of the convergence between datasciences and neurosciences. The production of intelligent agents that constitute learning machines modifies the meaning and function of being intelligent agents and indeed being thinking subjects. The imaginary of intelligent agents functions with the illusory of autonomous agents that can act on their learning.

If governing polities as learning machines produces a new kind of subject that is neither autonomic nor autonomous – the autopoietic subject – then we need to understand how this changes the citizenship game. What exactly is a citizenship game? It is a game that can only be played by autonomous citizen-subjects. The citizenship game was a revolution against the kingship game around 2,500 years ago. The citizen-subject was an autonomous agent exercising power on others and themselves without the authority of a sovereign. Over the next 2,500 years, the multiple rises and falls of the citizen-subject shaped the way power was exercised, by whom and on whom, as well as the rights and obligations associated with it. If colonisation, slavery and wars were their lowest ebb 500 years ago, the revolutions against the kingship game were at their highest flow around 250 years ago. The population game, which emerged then, laid the groundwork for the rise of the citizen-subject as a 'modern' autonomous agent. The essential questions that now arise are: what kind of citizen-subject does the imitation game produce, and how are polities structured to enable its functioning? If polities have now emerged as learning machines by deterritorialising and reterritorialising polities such as cities, states and empires, what kind of citizen-subject do they give rise to?

We trace the concepts of autopoietic, sentient, platform and networked as operative concepts producing a new subject. Sensory power passes through subjects as sentient beings as combinations of artificial and natural beings and then incorporate or integrate them into platforms, with self-generating responses to signals, weaving networks of impulses. The practices of capturing subjects, monitoring their performance, detecting anomalies and repairing or correcting them by pulses make polities function as learning machines. This precipitates the emergence of ubiquitous methods such as continuous, simultaneous and instantaneous sensing, identifying, ranking and evaluating of pulses and signals. How does the autopoietic subject negotiate this labyrinth?

It is tempting to conclude this book, which is an invitation to think about power, especially sensory power, differently, with Italo Calvino's enigmatic but poignant lecture 'Cybernetics and Ghosts' (1967).[3] Calvino organises his lecture around the relationship between words and things over a very long period of time.[4] He paints an imaginative history of their relations as having evolved from narrow repertoires into complex patterns of storytelling that gave birth to literature as storytelling. Between words and things, the author emerges as an authorised storyteller that imposes their discrete view on the story. Until the twentieth century the author is not a function but a creator. Yet, Calvino says, the twentieth century is on the verge of forging a new relationship between words and things where the author is no longer a creator but a function of a linguistic apparatus of writing with its rules, grammar, style and structure.

Calvino says that it is not only structuralists that are taking language apart and putting it together but also scientists with their cybernetic machines. The author is being decentred by linguistics and cybernetics simultaneously. If linguistics moves away from a correspondence between words and things, cybernetics is similarly moving towards constituting words and things 'as a series of discontinuous states, of combinations of impulses acting on a finite (though enormous) number of sensory and motor organs' (Calvino [1980] 1997, Section I). Calvino asks whether learning machines ('electronic brains') are providing a new model of our memory, imagination and even conscience. Calvino has read his cybernetics authors and thinks that Shannon, Weiner, von Neumann and Turing were decentring the author with 'the rapid passage of signals on the intricate circuits that connect the relays, the diodes, the transistors with which our skulls are crammed'. Are not linguistics and cybernetics both talking about codes and messages and establishing an entropy of language? If the literature machine is a machine 'capable of learning, of changing their own programs, of developing their own sensibilities and their own needs, nothing prevents us from foreseeing a literature machine that at a certain point feels unsatisfied with its own traditionalism and starts to propose new ways of writing, turning its own codes completely upside down'. This is Calvino's view of the imitation game, where the author as creator loses it or becomes a machine.

Calvino says he suspects that the imitation game was being played for some time as he always questioned the theories of the author as creator of ideas out of nothing. This is because 'in these theories there always remained a void that no one knew how to fill, a zone of darkness between cause and effect: how does one arrive at the written page? By what route is the soul or history or society or the subconscious transformed into a series of black lines on a white page?' What these theories call 'genius or talent or inspiration or intuition is nothing other than finding the right road empirically, following one's nose, taking short cuts, whereas the machine would follow a systematic and conscientious route while being extremely rapid and multiple at the same time' (Calvino [1980] 1997, Section II).

Calvino asks if we now conceive of the brain as a complex network of interconnected circuits, relays, diodes and transistors within the body, through which signals rapidly pass, does our intelligence consists in coding and decoding these signals and acting on them? It is entirely possible 'with a clear conscience and without regrets that [the author] could perfectly well be occupied by a mechanical device' (Calvino 1997, Section III).

However, Calvino has another trick up his sleeve: what if creative or critical writing were games that only we can play? If literature is not merely about describing but rather about creating and bringing things into play, would there have been literature in the first place? Calvino says that 'literature follows paths that flank and cross the barriers of prohibition, that leads to saying what could not be said, to an invention that is always a reinvention of words and stories that have been banished from the individual or collective memory'. So 'the literature machine can perform all the permutations possible on a given material, but the poetic result will be the particular effect of one of these permutations on a [writer] endowed with a consciousness and an unconscious, that is, an empirical and historical [writer]'. Calvino thinks that the literature machine can produce authors but playfully questions whether they can become poets.

Calvino finds this question unanswerable but offers an allegory.[5] This is the story of two prisoners who devise a plan to escape their captivity. One prisoner attempts to construct a labyrinth of tunnels within the prison machine, while the other creates a precise blueprint of the machine.[6] Calvino concludes that to comprehend how the machine was constructed one must not only dig a labyrinth but also create a blueprint.[7]

Although we are confronted today with as difficult a game as Calvino's allegory expresses, how polities as learning machines produced by sensory power organises life is perhaps different from Calvino's modified prisoner's dilemma, played by Nietzsche's Apollo and Dionysius. We find ourselves in a game where the kingship game (doing things with swords), citizenship game (doing things with words) and population game (doing things with numbers) are conjoined by the imitation game (doing things with neurons) and are all at play.

Throughout this book, we have repeatedly emphasised how projects and professions, as the driving forces behind sensory power, are reshaping sovereign, disciplinary and regulatory forms of power. We have frequently reminded ourselves that the accumulation of capital and the accumulation of subject peoples are closely intertwined. The scale and scope of projects that accumulate various forms of capital, such as constructing entirely new cities as learning machines, constructing immense infrastructures for data storage with insatiable energy demands (which have led corporations to develop their own nuclear energy programmes), construct transmission networks ranging from ubiquitous masts and undersea fibre networks to satellites orbiting the planet earth, and extract and exploit new minerals for the production of microchips and sensors, are manifestations of sensory power. These manifestations all reflect the interconnected games in which we find ourselves, which accumulate new subjects. The advertiser, broker, follower, hacker, homeguest, homehost, influencer, investor (e.g. fintech, proptech, agrotech), startup, streamer and subscriber are just a few but growing names of subjects accumulated by the strategies and technologies of sensory power, with effects and affects as profound and extensive as anything we have experienced as humans. Confronting sensory power and playing the imitation game differently necessitates both unravelling labyrinths and creating blueprints of how the learning machines operate and how the imitation game is played and how it is changing kingship, citizenship and population games. This book cannot claim to have either unravelled labyrinths or created blueprints of the situation in which we find ourselves. Instead, it has attempted to show how polities as learning machines play the imitation game in the context of the long history of power and the autopoietic subject it has produced.

Notes

1. It is important to articulate such movements as datapolitics (Bigo, Isin, and Ruppert 2019) and neuropolitics (Connolly 2002) but, for reasons we explain below, 'what games shall we play?' is a broader question of strategy as a countermovement to sensory power. Could we imagine

biopoetics as a countermovement to biopolitics? After struggling with the concept of biopower (which we call regulative power) and its associative politics as biopolitics (which we call population game), Foucault once, and only once, offers biopoetics as a counteractivity to biopolitics that would cultivate and recognise such relational rights. According to the editors, Foucault does not say during the lecture but offer the following explanation for the concepts in the manuscript: 'biopoetics would be justified because it is indeed a sort of personal fabrication of one's own life (note that in these arts the question often arises of whether or not an act is beautiful). One could thus follow the problem of sexual conduct: biopoetics where it is a matter of the aesthetic-moral conduct of individual existence; biopolitics where it is a matter of the normalization of sexual conducts according to what is considered politically as the requirement of a population' (Foucault [1980–1981] 2017, 34).

2. By approaching a long history of power, we distinguish between power (in) theory and power (in) practice. The former is its essence while the latter is its historic forms of existence. What are these forms? How to do things with sovereign power, how to do things with disciplinary power and how to do things with regulative power are empirical questions and addressing them involves not only a reworking of Spinoza, Nietzsche, Deleuze and Foucault on power but also supplementing it with Austin and Derrida and being guided by Bourdieu.
3. The lecture 'Cybernetics and Ghosts' was delivered in Turin in 1967 and was published in a collection of essays as *The Literature Machine* (Calvino [1980] 1997).
4. Incidentally, Foucault's *Les mots et les choses* (words and things) was published in 1966 and translated into English as *The Order of Things* (Foucault [1966] 2002). Calvino does not explicitly mention Foucault but his thoughts on the author-function seem both critical of and inspired by Foucault, whom he probably considers as among 'French structuralists' when he discusses Roland Barthes and Tel Quel. Calvino says: 'I will not go into technical details on which I could only be an unauthorized and rather unreliable commentator. My intention here is merely to sum up the situation, to make connections between a number of books I have recently read, and to put these in the context of a few general reflections' (Calvino [1980] 1997, Section I).
5. To articulate this problem, Calvino mentions his recent short story, which retells Alexandre Dumas's *Le Comte de Monte-Cristo*. Calvino does not say this but it is hard to escape the conclusion that these prisoners are Nietzsche's Apolline and Dionysiac figures. The tension between an

inductive (digging tunnels in the machine) and a deductive (imagining the machine) approach to escape resonates with Nietzsche. 'The labyrinth is made so that whoever enters it will stray and get lost. But the labyrinth also poses the [prisoner] a challenge: that he reconstructs the plan of it and dissolve its power. If he succeeds, he will have destroyed the labyrinth; for one who has passed through it, no labyrinth exists' (Calvino [1980] 1997, Section VI).

6. He concludes both the lecture and the story with the following. 'If I succeed in mentally constructing a fortress from which it is impossible to escape, this imagined fortress either will be the same as the real one—and in this case it is certain we shall never escape from here, but at least we will achieve the serenity of knowing we are here because we could be nowhere else – or it will be a fortress from which escape is even more impossible than from here – which would be a sign that here an opportunity of escape exists: we have only to identify the point where the imagined fortress does not coincide with the real one and then find it' (Calvino [1980] 1997, Section VI).
7. A few scholars have discussed this lecture. Martin McLaughlin (1998) discusses it within the context of Calvino's broader work especially in relation to Calvino's (1972) *Invisible Cities*. Franco Ricci (2006) discusses Calvino's works as articulating and playing difficult games.

References

Abraham, Anuj, Chetan B. Math, Shitala Prasad, and Mohit Sharma. 2024. 'Real-Time Traffic Accident Detection for an Intelligent Mobility in Smart Cities.' In *Internet of Everything for Smart City and Smart Healthcare Applications*, edited by Nishu Gupta and Sumita Mishra, 145–61. Cham: Springer.

Adli, Mazda, Maximilian Berger, Eva-Lotta Brakemeier, Ludwig Engel, Joerg Fingerhut, Ana Gomez-Carrillo, Rainer Hehl, et al. 2017. 'Neurourbanism: Towards a New Discipline'. *The Lancet Psychiatry* 4 (3): 183–85.

Agamben, Giorgio. 1998. *Homo Sacer: Sovereign Power and Bare Life*. Stanford, CA: Stanford University Press.

Agar, Jon. 2003. *The Government Machine: A Revolutionary History of the Computer.* Cambridge, MA: MIT Press.

Alavi, Amir H., Maria Q. Feng, Pengcheng Jiao, and Zahra Sharif-Khodaei. 2022. 'Advanced Sensing and Monitoring Systems for Smart Cities.' In *The Rise of Smart Cities*, edited by Amir H. Alavi, Maria Q. Feng, Pengcheng Jiao, and Z. Sharif Khodaei, 1–26. Amsterdam: Elsevier.

Albino, Vito, Umberto Berardi, and Rosa Maria Dangelico. 2015. 'Smart Cities: Definitions, Dimensions, Performance, and Initiatives.' *Journal of Urban Technology* 22 (1): 3–21.

Amaxilatis, Dimitrios, Georgios Mylonas, Evangelos Theodoridis, Luis Diez, and Katerina Deligiannidou. 2020. 'LearningCity: Knowledge Generation for Smart Cities.' In *Smart Cities Performability, Cognition, & Security*, edited by Fadi Al-Turjman, 17–41. Cham: Springer.

Ambika, D. R., K. R. Radhika, and D. Seshachalam. 2020. 'Identity Verification Using Biometrics in Smart-Cities.' In *Smart Cities Performability, Cognition & Security*, edited by Fadi Al-Turjman, 169–99. Cham: Springer.

Amoore, Louise. 2015. *Algorithmic Life: Calculative Devices in the Age of Big Data*. Abingdon: Routledge.

Anderson, Margo J. 2015. *The American Census: A Social History*. 2nd ed. New Haven, CT: Yale University Press.

Angelidou, Margarita. 2015. 'Smart Cities: A Conjuncture of Four Forces.' *Cities* 47: 95–106.

Arendt, Hannah. (1955) 2005. 'Introduction into Politics.' In *The Promise of Politics*, edited by Jerome Kohn, 93–200. New York: Schocken Books.

Arendt, Hannah. (1958) 2018. *The Human Condition*. 2nd ed. Chicago: University of Chicago Press.

Austin, J. L. 1962. *How to Do Things with Words*. Edited by J. O. Urmson. Oxford: Oxford University Press.

Austin, J. L. (1961) 1970. *Philosophical Papers*. Edited by J. O. Urmson and G. J. Warnock. 2nd ed. Oxford: Oxford University Press.

Aydın, Can. 2022. 'Conceptual Design: Components of Smart Cities.' In *Artificial Intelligence Perspective for Smart Cities*, edited by Sezer Bozkuş Kahyaoğlu and Vahap Tecim, 7–18. Boca Raton, FL: CRC Press.

Baker, David. 2019. 'Against the Grain: A Deep History of the Earliest States.' *Cambridge Review of International Affairs* 32 (2): 208–12.

Barker, Graeme, and Candice Goucher, eds. 2015. *A World with Agriculture, 12,000 BCE–500 CE*. Vol. 2. The Cambridge World History. Cambridge: Cambridge University Press.

Batty, M., K. W. Axhausen, F. Giannotti, A. Pozdnoukhov, A. Bazzani, M. Wachowicz, G. Ouzounis, and Y. Portugali. 2012. 'Smart Cities of the Future.' *European Physical Journal-Special Topics* 214 (1): 481–518.

Bauman, Z., D. Bigo, P. Esteves, E. Guild, V. Jabri, D. Lyon, and R. B. J. Walker. 2014. 'After Snowden: Rethinking the Impact of Surveillance.' *International Political Sociology* 8 (2): 121–44.

Beckman, Frida, ed. 2020. *Control Culture: Foucault and Deleuze after Discipline*. Edinburgh: Edinburgh University Press.

Benzécri, J.-P. (1973) 1992. *Correspondence Analysis Handbook*. Translated by T. K. Gopala. New York: CRC Press.

Bernhardt, Chris. 2016. *Turing's Vision: The Birth of Computer Science*. Cambridge, MA: MIT Press.

Bielza, Concha, and Pedro Larrañaga. 2020. *Data-Driven Computational Neuroscience: Machine Learning and Statistical Models*. Cambridge: Cambridge University Press.

Bigo, Didier, Engin Isin, and Evelyn Ruppert, eds. 2019. *Data Politics: Worlds, Subjects, Rights*. London: Routledge.

Boden, Margaret A. 2006. *Mind as Machine: A History of Cognitive Science.* 2 vols. Oxford: Oxford University Press.

Boden, Margaret A. 2016. *AI: Its Nature and Future.* Oxford: Oxford University Press.

Boltanski, Luc. 2016. 'Alain Desrosières's Reflexive Numbers.' In *The Social Sciences of Quantification*, edited by Isabelle Bruno, Florence Jany-Catrice, and Béatrice Touchelay, 33–42. Logic, Argumentation & Reasoning. Cham: Springer International Publishing.

Bourdieu, Pierre. (1979) 1984. *Distinction: A Social Critique of the Judgement of Taste.* Stanford, CA: Harvard University Press.

Bourdieu, Pierre. (1984) 1988. *Homo Academicus.* Stanford, CA: Stanford University Press.

Bourdieu, Pierre. 1991. *Language and Symbolic Power.* Cambridge, MA: Harvard University Press.

Bourdieu, Pierre. (2015) 2018. *Classification Struggles, Lectures at the Collège de France (1981–1982).* Translated by Peter Collier. Cambridge: Polity.

Bower, James M, ed. 2013. *20 Years of Computational Neuroscience.* New York, NY: Springer New York.

Boyer, M. Christine. (1983) 1994. *Dreaming the Rational City: The Myth of American City Planning.* 2nd ed. Cambridge, MA: MIT Press.

Boyer, M. Christine. 1996. *Cybercities: Visual Perception in the Age of Electronic Communication.* New York: Princeton Architectural Press.

Bulmer, Michael. 2003. *Francis Galton: Pioneer of Heredity and Biometry.* Baltimore, MD: John Hopkins University Press.

Burns, Robert. 1993. *Selected Poems.* Edited by Carol McGuirk. London: Penguin.

Butler, Judith. 1990. *Gender Trouble: Feminism and the Subversion of Identity.* London: Routledge.

Butler, Judith. 2004. *Undoing Gender.* London: Routledge.

Butler, Judith. (1997) 2021. *Excitable Speech: A Politics of the Performative.* 2nd edition. London: Routledge.

Calvino, Italo. 1972. *Invisible Cities.* London: Picador.

Calvino, Italo. (1980) 1997. *The Literature Machine: Essays.* Translated by Patrick Creagh. London: Vintage.

Cao, Longbing. 2018. 'Data Science: A Comprehensive Overview.' *ACM Computing Surveys* 50 (3): 1–42.

Capponi, Andrea, Claudio Fiandrino, Burak Kantarci, Luca Foschini, Dzmitry Kliazovich, and Pascal Bouvry. 2019. 'A Survey on Mobile Crowdsensing Systems: Challenges, Solutions, and Opportunities.' *IEEE Communications Surveys & Tutorials* 21 (3): 2419–65.

Cardone, Giuseppe, Andrea Cirri, Antonio Corradi, Luca Foschini, Raffaele Ianniello, and Rebecca Montanari. 2014. 'Crowdsensing in Urban Areas for City-Scale Mass Gathering Management: Geofencing and Activity Recognition.' *IEEE Sensors Journal* 14 (12): 4185–95.

Cardullo, Paolo, and Rob Kitchin. 2019. 'Smart Urbanism and Smart Citizenship: The Neoliberal Logic of "Citizen-Focused" Smart Cities in Europe.' *Environment and Planning C-Politics and Space* 37 (5): 813–30.

Castells, Manuel. 1989. *The Informational City: Information Technology, Economic Restructuring and the Urban Regional Process*. Oxford: Blackwell.

Castells, Manuel. (1996) 2011. *The Rise of the Network Society*. 2nd ed. Malden, MA: Wiley-Blackwell.

Cavell, Stanley. 1994. *A Pitch of Philosophy: Autobiographical Exercises*. Cambridge, MA: Harvard University Press.

Changeux, Jean-Pierre, and Jean Chavaillon, eds. 1996. *Origins of the Human Brain*. Oxford University Press.

Christian, David, ed. 2015. *Introducing World History, to 10,000 BCE*. Vol. 1. The Cambridge World History. Cambridge: Cambridge University Press.

Church, Alonzo. 1936. 'A Note on the Entscheidungsproblem.' *Journal of Symbolic Logic* 1 (1): 40–41.

Churchland, Patricia Smith, and Terrence J. Sejnowski. (1992) 2017. *The Computational Brain*. 25th Anniversary Edition. Cambridge, MA: MIT Press.

Citton, Yves; 2016. *The Ecology of Attention*. Cambridge: Polity Press.

Clark, Peter, ed. 2013. *The Oxford Handbook of Cities in World History*. Oxford: Oxford University Press.

Cleveland, William S. 2014. 'Data Science: An Action Plan for Expanding the Technical Areas of the Field of Statistics.' *Statistical Analysis and Data Mining: The ASA Data Science Journal* 7 (6): 414–17.

Cockerham, William C., and Brian P. Hinote. 2009. 'Quantifying Habitus: Future Directions.' In *Quantifying Theory: Pierre Bourdieu*, edited by Chris Sanders and Karen Robson. New York: Springer.

Coeckelbergh, Mark. 2022. *Self-Improvement: Technologies of the Soul in the Age of Artificial Intelligence*. No Limits. New York: Columbia University Press.

Coelho, Vitor N., Thays A. Oliveira, Iara V. O. Figueiredo, Marcone J. F. Souza, and Iuri Veloso. 2019. 'A Multicriteria View About Judicial and Legislative Decision Making in Digital Cities and Societies.' In *Smart and Digital Cities*, edited by Vitor Nazário Coelho, Igor Machado Coelho, Thays A.Oliveira, and Luiz Satoru Ochi, 209–20. Cham: Springer.

Connolly, William E. 2002. *Neuropolitics: Thinking, Culture, Speed*. Minneapolis: University of Minnesota Press.

Conway, Flo, and Jim Siegelman. 2005. *Dark Hero of the Information Age: In Search of Norbert Wiener, the Father of Cybernetics*. New York, NY: Basic Books.

Cordery, Simon. 2003. *British Friendly Societies, 1750–1914*. London: Palgrave Macmillan.

Coyne, Richard. 2023. *Cryptographic City*. Cambridge, MA: MIT Press.

Cremonini, Marco. 2024. *Data Science Fundamentals with R, Python, and Open Data*. Hoboken, NJ: Wiley.

Cugurullo, Federico. 2021. *Frankenstein Urbanism: Eco, Smart and Autonomous Cities, Artificial Intelligence and the End of the City*. London: Routledge.

Cunha, David. 2021. 'Creating a Smart Ecosystem in Lisbon.' In *How Smart Is Your City?* edited by Maria Isabel Aldinhas Ferreira, 145–60. Cham: Springer.

Curtis, Bruce. 2001. *The Politics of Population: State Formation, Statistics, and the Census of Canada, 1840–1875*. Toronto: University of Toronto Press.

Damasio, Antonio. 2006. *Descartes' Error: Emotion, Reason and the Human Brain*. New York, NY: Vintage.

Datta, Ayona. 2018. 'The Digital Turn in Postcolonial Urbanism: Smart Citizenship in the Making of India's 100 Smart Cities.' *Transactions of the Institute of British Geographers* 43 (3): 405–19.

De Brouwer, Philippe J. S. 2020. *The Big R-Book: From Data Science to Learning Machines for the Professional*. Hoboken, NJ: Wiley.

De Castro Neto, Miguel, and Tiago De Melo Cartaxo. 2021. 'Algorithmic Cities: A Dystopic or Utopic Future?' In *How Smart Is Your City?* edited by Maria Isabel Aldinhas Ferreira, 59–73. Cham: Springer.

Deleuze, Gilles. (1962) 1985. *Nietzsche and Philosophy*. Translated by Hugh Tomlinson. New York: Columbia University Press.

Deleuze, Gilles. (1986) 1988. *Foucault*. Translated by Seán Hand. Minneapolis: University of Minnesota Press.

Deleuze, Gilles. (1969) 1988. *Spinoza: Practical Philosophy*. London: City Lights Books.

Deleuze, Gilles. (1990) 1995a. 'Control and Becoming.' In *Negotiations, 1972–1990*, translated by Martin Joughin, 169–76. New York: Columbia University Press.

Deleuze, Gilles. (1990) 1995b. 'PostScript on Control Societies.' In *Negotiations, 1972–1990*, translated by Martin Joughin, 177–82. New York: Columbia University Press.

Deleuze, Gilles, and Felix Guattari. (1980) 1987. *A Thousand Plateaus: Capitalism and Schizophrenia*. Translated by Brian Massumi. Minneapolis: University of Minnesota Press.

Derrida, Jacques. (1977) 1988. *Limited Inc*. Edited by Gerald Graff. Evanston, IL: Northwestern University Press.

Desrosières, Alain. 1991. 'The Part in Relation to the Whole: How to Generalise? The Prehistory of Representative Sampling.' In *The Social Survey in Historical Perspective: 1880–1940*, edited by Martin Bulmer, Kevin Bales, and Kathryn Kish Sklar, 217–44. Cambridge: Cambridge University Press.

Desrosières, Alain. (1993) 1998. *The Politics of Large Numbers: A History of Statistical Reasoning*. Translated by Camille Naish. Cambridge, MA: Harvard University Press.

Desrosières, Alain. 2014. *Prouver et Gouverner. Une Analyse Politique Des Statistiques Publiques*. Paris: La Découverte.

Desrosières, Alain. 2016. 'The Quantification of the Social Sciences: An Historical Comparison.' In *The Social Sciences of Quantification From Politics of Large Numbers to Target-Driven Policies*, edited by Isabelle Bruno, Florence Jany-Catrice, and Béatrice Touchelay, 183–204. Berlin: Springer.

Deutsch, Karl W. (1963) 1966. *The Nerves of Government: Models of Political Communication and Control*. 2nd ed. New York: Free Press.

Doherty, Paul. 2023. *Smart Cities: Reimagining the Urban Experience*. Milwaukee, WI: Quality Press.

Dreyfus, Hubert L. 1972. *What Computers Can't Do: The Limits of Artificial Intelligence*. New York: Harper & Row.

Dreyfus, Hubert L. (1979) 1992. *What Computers Still Can't Do: A Critique of Artificial Reason*. Cambridge, MA: MIT Press.

Dreyfus, Hubert L., and Stuart E. Dreyfus. 1986. *Mind over Machine: The Power of Human Intuition and Expertise in the Era of the Computer*. New York: Free Press.

Dyer-Witheford, Nick, Atle Mikkola Kjøsen, and James Steinhoff. 2019. *Inhuman Power: Artificial Intelligence and the Future of Capitalism*. London: Pluto Press.

Dyson, George. 2012. *Turing's Cathedral: The Origins of the Digital Universe*. New York: Pantheon.

Eberhard, John P. 2009. *Brain Landscape: The Coexistence of Neuroscience and Architecture*. Oxford: Oxford University Press.

Elden, Stuart. 2016. *Foucault's Last Decade*. Cambridge: Polity.

Elden, Stuart. 2017. *Foucault: The Birth of Power*. Cambridge: Polity.

Emigh, Rebecca Jean, Dylan Riley, and Patricia Ahmed. 2016a. *Antecedents of Censuses from Medieval to Nation States*. Houndmills: Palgrave Macmillan.

Emigh, Rebecca Jean, Dylan Riley, and Patricia Ahmed. 2016b. *Changes in Censuses from Imperialist to Welfare States*. New York: Palgrave Macmillan US.

Esposito, Roberto. 2008. *Bios: Biopolitics and Philosophy*. Translated by Timothy Campbell. *Bi\0301os. English*. Minneapolis: University of Minnesota Press.

Felman, Shoshana. (1980) 2003. *The Scandal of the Speaking Body: Don Juan with J. L. Austin, or Seduction in Two Languages*. Translated by Catherine Porter. Stanford, CA: Stanford University Press.

Few, Stephen. 2006. *Information Dashboard Design: The Effective Visual Communication of Data*. London: O'Reilly.

Fiandrino, Claudio, Andrea Capponi, Giuseppe Cacciatore, Dzmitry Kliazovich, Ulrich Sorger, Pascal Bouvry, Burak Kantarci, Fabrizio Granelli, and Stefano Giordano. 2017. 'CrowdSenSim: A Simulation Platform for Mobile Crowdsensing in Realistic Urban Environments.' *IEEE Access* 5: 3490–503.

Fitzgerald, Des, Nikolas Rose, and Ilina Singh. 2016. 'Revitalizing Sociology: Urban Life and Mental Illness between History and the Present.' *The British Journal of Sociology* 67 (1): 138–60.

Foucault, Michel. (1969) 1972. *The Archaeology of Knowledge and the Discourse on Language*. London: Tavistock.

Foucault, Michel. (1975) 1977. *Discipline and Punish: The Birth of the Prison*. Translated by Alan Sheridan. New York: Vintage.

Foucault, Michel. (1971) 1977. 'Nietzsche, Genealogy, History.' In *Language, Counter-Memory, Practice*, edited and translated by Donald F. Bouchard, 1–14. Ithaca, NY: Cornell University Press.

Foucault, Michel. (1976) 1978. *The History of Sexuality: An Introduction*. Translated by Robert Hurley. New York: Pantheon Books.

Foucault, Michel. 1979. *Power, Truth, Strategy*. Edited by Meaghan Morris and Paul Patton. Sydney: Feral Publications.

Foucault, Michel. (1977) 1980. 'Truth and Power.' In *Power/Knowledge*, edited by Colin Gordon, translated by Colin Gordon, Leo Marshall, John Mepham, and Kate Soper, 109–33. New York, NY: Pantheon.

Foucault, Michel. 1982. 'The Subject and Power.' *Critical Inquiry* 8 (4): 777–95.

Foucault, Michel. (1984) 1997. 'What Is Enlightenment?' In *Ethics: Subjectivity and Truth*, edited by Paul Rabinow, 1:303–19. Essential Works of Foucault, 1954–1984. New York: The New Press.

Foucault, Michel. (1982) 2000. 'The Subject and Power.' In *Power*, edited by J. D. Faubion. Essential Works of Foucault, 1954–1984. New York: The New Press.

Foucault, Michel. (1973) 2000. 'Truth and Juridical Forms.' In *Power*, edited by J. D. Faubion, 3:1–89. Essential Works of Foucault, 1954–1984. New York: The New Press.

Foucault, Michel. (1966) 2002. *The Order of Things: An Archaeology of the Human Sciences*. London: Routledge.

Foucault, Michel. (1997) 2003. *Society Must Be Defended: Lectures at the Collège De France, 1975–76*. Translated by D. Macey. New York: Picador.

Foucault, Michel. (2004) 2007. *Security, Territory, Population: Lectures at the Collège de France 1977–1978*. Translated by Graham Burchell. Basingstoke: Palgrave Macmillan.

Foucault, Michel. (2004) 2008. *The Birth of Biopolitics: Lectures at the Collège de France, 1978–1979*. Translated by Graham Burchell. Basingstoke: Palgrave Macmillan.

Foucault, Michel. (2011) 2013. *Lectures on the Will to Know: Lectures at the Collège De France, 1970–1971; and Oedipal Knowledge*. Edited by Daniel Defert. Translated by Graham Burchell. Basingstoke: Palgrave Macmillan.

Foucault, Michel. 2016. *The Punitive Society: Lectures at the Collège De France, 1972–1973*. Translated by Graham Burchell. Basingstoke: Palgrave Macmillan.

Foucault, Michel. (1980–1981) 2017. *Subjectivity and Truth: Lectures at the Collège de France, 1980–1981*. Edited by Arnold Davidson. Translated by Graham Burchell. London: Palgrave Macmillan.

Foucault, Michel. 2019a. *Discourse and Truth & Parrēsia*. Chicago: Chicago University Press.

Foucault, Michel. 2019b. *Penal Theories and Institutions: Lectures at the Collège de France, 1971–1972*. Chicago: University of Chicago Press.

Frapporti, Mattia. 2024. 'The Politics of Platforms. Exploring Platforms' Infrastructural Role and Power.' In *Capitalism in the Platform Age: Emerging Assemblages of Labour and Welfare in Urban Spaces*, edited by Sandro Mezzadra, Niccolò Cuppini, Mattia Frapporti, and Maurilio Pirone, 81–95. Cham: Springer.

Freud, Sigmund. 2001. *The Complete Psychological Works of Sigmund Freud: 'The Future of An Illusion', 'Civilization and Its Discontents' and Other Works Vol 21*. New York: Vintage.

Gabrys, Jennifer. 2014. 'Programming Environments: Environmentality and Citizen Sensing in the Smart City.' *Environment and Planning D: Society and Space* 32 (1): 30–48.

Gabrys, Jennifer. 2019. 'Data Citizens: How to Reinvent Rights.' In *Data Politics: Worlds, Subjects, Rights*, edited by Didier Bigo, Engin Isin, and Evelyn Ruppert, 249–66. London: Routledge.

Galloway, Alexander. 2006. *Protocol: How Control Exists after Decentralization*. Cambridge, MA: MIT Press.

Ganti, Raghu, Fan Ye, and Hui Lei. 2011. 'Mobile Crowdsensing: Current State and Future Challenges.' *IEEE Communications Magazine* 49 (11): 32–39.

Georgiou, Myria. 2024. *Being Human in Digital Cities*. Cambridge: Polity.

Glass, D. V. 1973. *Numbering the People. The Eighteenth-Century Population Controversy and the Development of Census and Vital Statistics in Britain*. Farnborough: Saxon House.

Godsey, Brian. 2017. *Think Like a Data Scientist: Tackle the Data Science Process Step-by-Step*. Shelter Island: Manning.

Gonçalves, Bernardo. 2023. *The Turing Test Argument*. New York: Routledge.

Goyal, Mukta, Ekata Gupta, Abhishek Srivastava, and Prabha Kiran. 2024. 'Smart Cities in the Age of AI and the Internet of Things.' In *Handbook of Artificial Intelligence for Smart City Development*, by Sandhya Makkar, Gobinath Ravindran, Ripon Kumar Chakrabortty, and Arindam Pal, 27–47. Boca Raton, FL: CRC Press.

Grommé, Francisca, Evelyn Ruppert, and Baki Cakici. 2018. 'Data Scientists: A New Faction of the Transnational Field of Statistics.' In *Ethnography for a Data-Saturated World*, edited by Hannah Knox and Dawn Nafus, 33–61. Materialising the Digital. Manchester: Manchester University Press.

Gros, Frédéric. 2015. 'Is There a Biopolitical Subject? Foucault and the Birth of Biopolitics.' In *Biopower: Foucault and Beyond*, edited by Vernon W. Cisney and Nicolae Morar, 259–73. Chicago: University of Chicago Press.

Grossi, Giuseppe, and Daniela Pianezzi. 2017. 'Smart Cities: Utopia or Neoliberal Ideology?' *Cities* 69: 79–85.

Hacking, Ian. 1981. 'How Should We Do the History of Statistics?' *Ideology & Consciousness* 8: 15–26.

Hacking, Ian. 1983. 'Biopower and the Avalanche of Printed Numbers.' *Humanities and Society* 5: 279–95.

Hacking, Ian. 1990. *The Taming of Chance*. Cambridge: Cambridge University Press.

Hacking, Ian. 2002. *Historical Ontology*. Cambridge, MA: Harvard University Press.

Hacking, Ian. (1975) 2006. *The Emergence of Probability: A Philosophical Study of Early Ideas About Probability, Induction and Statistical Inference*. 2nd ed. Cambridge: Cambridge University Press.

Hacking, Ian. 2007. 'Kinds of People: Moving Targets.' *Proceedings of the British Academy* 151: 285–318.

Haigh, Thomas, and Paul E. Ceruzzi. (1998) 2021. *A New History of Modern Computing*. 3rd ed. History of Computing. Cambridge, MA: MIT Press.

Hald, Anders. 2005. *A History of Probability and Statistics and Their Applications before 1750*. New York, NY: John Wiley & Sons.

Halegoua, Germaine R. 2019. *The Digital City: Media and the Social Production of Place*. Critical Cultural Communication. New York: New York University Press.

Hardt, Michael, and Antonio Negri. 2000. *Empire*. Cambridge: Harvard University Press.

Hassabis, Demis, Dharshan Kumaran, Christopher Summerfield, and Matthew Botvinick. 2017. 'Neuroscience-Inspired Artificial Intelligence.' *Neuron* 95 (2): 245–58.

Heims, Steve. 1980. *John von Neumann and Norbert Wiener: From Mathematics to the Technologies of Life and Death*. Cambridge, MA: MIT Press.

Hennig, Christian, and Marina Meila. 2016. 'Cluster Analysis: An Overview.' In *Handbook of Cluster Analysis*, edited by Christian Hennig, Marina Meila, Fionn Murtagh, and Roberto Rocci, 1–19. London: CRC Press.

Higgs, Edward. 2004. *The Information State in England: The Central Collection of Information on Citizens, 1500–2000*. Basingstoke: Palgrave Macmillan.

Hudson-Smith, Andrew, Stephan Hügel, and Flora Roumpani. 2020. 'Self-Monitoring, Analysis and Reporting Technologies: Smart Cities and Real-Time Data.' In *The Routledge Companion to Smart Cities*, edited by Katharine S. Willis and Alessandro Aurigi, 383–94. London: Routledge.

Isin, Engin. 2002. *Being Political: Genealogies of Citizenship*. Minneapolis: University of Minnesota Press.

Isin, Engin. 2004. 'The Neurotic Citizen.' *Citizenship Studies* 8 (3): 217–35.

Isin, Engin. 2024. *Citizenship*. New Trajectories in Law. London: Routledge.

Isin, Engin, and Evelyn Ruppert. 2019. 'Data's Empire: Postcolonial Data Politics.' In *Data Politics: Worlds, Subjects, Rights*, edited by Didier Bigo, Engin Isin, and Evelyn Ruppert. Routledge Studies in International Political Sociology. London: Routledge.

Isin, Engin, and Evelyn Ruppert. 2020a. *Being Digital Citizens*. 2nd ed. London: Rowman & Littlefield.

Isin, Engin, and Evelyn Ruppert. 2020b. 'The Birth of Sensory Power: How a Pandemic Made It Visible.' *Big Data & Society*, 1–15.

Jain, Anil K. 2010. 'Data Clustering: 50 Years beyond K-Means.' *Pattern Recognition Letters* 31 (8): 651–66.

Jain, Anil K., M. N. Murty, and P. J. Flynn. 1999. 'Data Clustering: A Review.' *ACM Computing Surveys* 31 (3): 264–323.

Jain, Anil K., A. Topchy, M. H. C. Law, and J. M. Buhmann. 2004. 'Landscape of Clustering Algorithms.' *Proceedings of the 17th International Conference on Pattern Recognition* ICPR 2004: 260–63.

Johns, Fleur. 2017. 'Data Territories: Changing Architectures of Association in International Law.' In *Netherlands Yearbook of International Law*, edited by M. Kuijer and W. Werner, 107–29.

Johns, Fleur. 2021. 'Governance by Data.' *Annual Review of Law and Social Science* 17 (1): 53–71.

Johns, Fleur, and Caroline Compton. 2022. 'Data Jurisdictions and Rival Regimes of Algorithmic Regulation.' *Regulation & Governance* 16 (1): 63–84.

Karvonen, Andrew, Federico Cugurullo, and Federico Caprotti, eds. 2019a. *Inside Smart Cities: Place, Politics and Urban Innovation*. London: Routledge.

Karvonen, Andrew, Federico Cugurullo, and Federico Caprotti. 2019b. 'The Long and Unsettled Future of Smart Cities.' In *Inside Smart Cities: Place, Politics and Urban Innovation*, edited by Andrew Karvonen, Federico Cugurullo, and Federico Caprotti, 291–98. London: Routledge.

Kertzer, David I. 2002. *Census and Identity: The Politics of Race, Ethnicity, and Language in National Censuses*. Cambridge University Press.

Kirwan, Christopher G., and Stefan V. Dobrev. 2022. 'Cities as Convergent Autopoietic Systems.' In *Artificial Intelligence, Machine Learning, and Optimization Tools for Smart Cities*, edited by Panos M. Pardalos, Stamatina Th. Rassia, and Arsenios Tsokas, 1–26. Cham: Springer.

Kitchin, Rob. 2014. *The Data Revolution: Big Data, Open Data, Data Infrastructures and Their Consequences*. London: SAGE.

Kitchin, Rob. 2021. *Data Lives: How Data Are Made and Shape Our World*.

Kitchin, Rob, Claudio Coletta, and Gavin McArdle. 2020. 'Governmentality and Urban Control.' In *The Routledge Companion to Smart Cities*, edited by Katharine S. Willis and Alessandro Aurigi, 109–22. London: Routledge.

Kitchin, Rob, Tracey P. Lauriault, and Gavin McArdle. 2015a. 'Knowing and Governing Cities through Urban Indicators, City Benchmarking and Real-Time Dashboards.' *Regional Studies, Regional Science* 2 (1): 6–28.

Kitchin, Rob, Tracey P. Lauriault, and Gavin McArdle. 2015b. 'Urban Indicators and Dashboards: Epistemology, Contradictions and Power/Knowledge.' *Regional Studies, Regional Science* 2 (1): 43–45.

Kitchin, Rob, Tracey P. Lauriault, and Gavin McArdle, eds. 2018. *Data and the City*. Abingdon: Routledge.

Kitchin, Rob, and Gavin McArdle. 2018. 'Urban Data and City Dashboards: Six Key Issues.' In *Data and the City*, edited by Rob Kitchin, Tracey P. Lauriault, and Gavin McArdle, 111–26. Abingdon: Routledge.

Kong, Linghe, Bowen Wang, and Guihai Chen. 2019. *When Compressive Sensing Meets Mobile Crowdsensing*. Singapore: Springer Singapore.

Kummitha, Rama Krishna Reddy, and Nathalie Crutzen. 2017. 'How Do We Understand Smart Cities? An Evolutionary Perspective.' *Cities* 67 (July): 43–52.

Kurzweil, Ray. 2006. *The Singularity Is Near: When Humans Transcend Biology*. London: Penguin.

Kurzweil, Ray. 2012. 'Foreword.' In *The Computer and the Brain*, edited by John von Neumann. 3rd ed. New Haven, CT: Yale University Press.

Landahl, Gustaf. 2021. 'Stockholm: Smart City.' In *How Smart Is Your City?* edited by Maria Isabel Aldinhas Ferreira, 127–44. Cham: Springer.

Latour, Bruno. 2000. 'The Berlin Key or How to Do Words with Things.' In *Matter, Materiality and Modern Culture*, edited by Paul Graves-Brown, 10–21. London: Routledge.

Latour, Bruno. 2004. 'Why Has Critique Run out of Steam? From Matters of Fact to Matters of Concern.' *Critical Inquiry* 30 (2): 225–48.

Le Roux, Brigitte, and Henry Rouanet. 2010. *Multiple Correspondence Analysis*. Quantitative Applications in the Social Sciences 163. Thousand Oaks, CA: Sage.

Lebaron, Frédéric, and Karen Robson. 2009. 'How Bourdieu "Quantified" Bourdieu: The Geometric Modelling of Data.' In *Quantifying Theory: Pierre Bourdieu*, edited by Karen Robson and Chris Sanders. New York: Springer.

LeDoux, Joseph E. 2002. *Synaptic Self: How Our Brains Become Who We Are*. London: Macmillan.

Lee, Jung Hoon, Marguerite Gong Hancock, and Mei-Chih Hu. 2014. 'Towards an Effective Framework for Building Smart Cities: Lessons from Seoul and San Francisco.' *Technological Forecasting & Social Change* 89: 80–99.

Leeuwen, Marco Henk Dammes van. 2016. *Mutual Insurance 1550–2015: From Guild Welfare and Friendly Societies to Contemporary Micro-Insurers*. Palgrave Studies in the History of Finance. London: Palgrave Macmillan.

Lemke, Thomas. 2011. *Biopolitics: An Advanced Introduction*. Translated by Eric Frederick Trump. New York: New York University Press.

Lemke, Thomas. 2019. *Foucault's Analysis of Modern Governmentality: A Critique of Political Reason*. Translated by Erik Butler. London: Verso.

Lemke, Thomas. 2021. *The Government of Things: Foucault and the New Materialisms*. New York, NY: New York University Press.

Lepakshi, Vijayalakshmi A., and Prashanth C. Ranga. 2024. 'Artificial Intelligence for Smart City Development.' In *Handbook of Artificial Intelligence for Smart City Development*, edited by Sandhya Makkar, Gobinath Ravindran, Ripon Kumar Chakrabortty, and Arindam Pal, 60–75. Boca Raton, FL: CRC Press.

Leppänen, Teemu, José Álvarez Lacasia, Yoshito Tobe, Kaoru Sezaki, and Jukka Riekki. 2017. 'Mobile Crowdsensing with Mobile Agents.' *Autonomous Agents and Multi-Agent Systems* 31 (1): 1–35.

Levitan, Kathrin. 2011. *A Cultural History of the British Census: Envisioning the Multitude in the Nineteenth Century*. 1st ed. Palgrave Studies in Cultural and Intellectual History. New York: Palgrave Macmillan.

Logani, Manasvi, and Sandhya Makkar. 2024. 'Machine Learning for Smart City AI Systems.' In *Handbook of Artificial Intelligence for Smart City Development*, edited by Sandhya Makkar, Gobinath Ravindran, Ripon Kumar Chakrabortty, and Arindam Pal, 1–26. Boca Raton, FL: CRC Press.

Lovelace, Ada. 1843. 'Sketch of the Analytical Engine Invented by Charles Babbage.' In *Scientific Memoirs*, edited by Richard Taylor, 3:691–730. Selected from the Transactions of Foreign Academies of Science and Learned Societies, and from Foreign Journals. London: Richard and John E. Taylor.

Lucassen, Jan, Leo Lucassen, and Patrick Manning, eds. 2010. *Migration History in World History: Multidisciplinary Approaches*. Vol. 3. Studies in Global Social History. Leiden: Brill.

Luhmann, Niklas. (1997) 2012. *Theory of Society*. Translated by Rhodes Barrett. Vol. 1. Stanford, CA: Stanford University Press.

Luhmann, Niklas. (1997) 2013. *Theory of Society*. Translated by Rhodes Barrett. Vol. 2. Stanford, CA: Stanford University Press.

Lyon, David. 2014. 'Surveillance, Snowden, and Big Data: Capacities, Consequences, Critique.' *Big Data & Society* 1 (2): 1–13.

Lyon, David. 2018. *The Culture of Surveillance: Watching as a Way of Life*. Cambridge: Polity.

Lytton, William W. 2002. *From Computer to Brain: Foundations of Computational Neuroscience*. New York: Springer.

Mackenzie, Adrian. 2015. 'The Production of Prediction: What Does Machine Learning Want?' *European Journal of Cultural Studies* 18 (4–5): 429–45

Mackenzie, Adrian. 2017. *Machine Learners: Archaeology of a Data Practice*. Cambridge, MA: MIT Press.

Malabou, Catherine. (2004) 2008. *What Should We Do with Our Brain?* Translated by Sebastian Rand. New York, NY: Fordham University Press.

Malabou, Catherine. 2022. *Plasticity: The Promise of Explosion*. Edited by Tyler M. Williams. Edinburgh: Edinburgh University Press.

Malik, Kamal, Harsh Sadawarti, Moolchand Sharma, Umesh Gupta, and Prayag Tiwari, eds. 2023. *Computational Techniques in Neuroscience*. London: CRC Press.

Manning, Patrick, and Tiffany Trimmer. 2020. *Migration in World History*. 3rd ed. Themes in World History. New York: Routledge.

Marsal-Llacuna, Maria-Lluïsa, and Mark Evan Segal. 2016. 'The Intelligenter Method (i) for Making "Smarter" City Projects and Plans.' *Cities* 55 (June): 127–38.

Martin, Nathan D. 2009. 'Quantifying Social Class: A Latent Clustering Approach.' In *Quantifying Theory: Pierre Bourdieu*, edited by Karen Robson and Chris Sanders. New York: Springer.

Mattern, Shannon. 2015. 'Mission Control: A History of the Urban Dashboard.' March 2015.

Mattern, Shannon. 2021. *A City Is Not a Computer: Other Urban Intelligences*. Princeton, NJ: Princeton University Press.

Maturana, Humberto R., and Francisco J. Varela. 1980. *Autopoiesis and Cognition: The Realization of the Living*. Dordrecht: Springer.

Mayer-Schönberger, Viktor, and Kenneth Cukier. 2013. *Big Data: A Revolution That Will Transform How We Live, Work and Think*. London: John Murray.

Mbembe, Achille. 2019. *Necropolitics*. Translated by Steve Corcoran. Durham, NC: Duke University Press.

McCulloch, Warren S., and Walter Pitts. 1943. 'A Logical Calculus of the Ideas Immanent in Nervous Activity.' *Bulletin of Mathematical Biophysics* 5: 115–33.

McLaughlin, Martin. 1998. *Italo Calvino*. Edinburgh: Edinburgh University Press.

Mignolo, Walter. 2011. *The Darker Side of Western Modernity: Global Futures, Decolonial Options*. Duke University Press.

Miller, Paul. 2018. *An Introductory Course in Computational Neuroscience*. Computational Neuroscience. Cambridge, MA: MIT Press.

Mitchell, William J. 2003. *Me++: The Cyborg Self and the Networked City*. Cambridge, MA: MIT Press.

Mumford, Lewis. 1938. *The Culture of Cities*. San Diego, CA: Harcourt Brace Jovanovich.

Mumford, Lewis. 1961. *The City in History: Its Origins, Its Transformations, and Its Prospects*. New York: Harcourt Brace International.

Mumford, Lewis. (1966) 1967. *The Myth of the Machine: Techniques and Human Development*. New York: Harcourt.

Mumford, Lewis. (1964) 1970. *The Myth of the Machine: The Pentagon of Power*. New York: Harcourt.

Murtagh, Fionn. 2016. 'A Brief History of Cluster Analysis.' In *Handbook of Cluster Analysis*, edited by Christian Hennig, Marina Meila, Fionn Murtagh, and Roberto Rocci, 21–32. Chapman & Hall/CRC Handbooks of Modern Statistical Methods. London: CRC Press.

Murtagh, Fionn. 2018. *Data Science Foundations: Geometry and Topology of Complex Hierarchic Systems and Big Data Analytics*. Chapman & Hall/CRC Computer Science & Data Analysis Series. Boca Raton, FL, London and New York: CRC Press, Taylor & Francis Group.

Murtagh, Fionn. 2019a. *Correspondence Analysis and Data Coding with Java and R*. Foreword by J.-P. Benzécri. Computer Science and Data Analysis Series. London: CRC.

Murtagh, Fionn. 2019b. 'The Geometry and Topology of Data and Information for Analytics of Processes and Behaviours: Building on Bourdieu and Addressing New Societal Challenges.' In *Empirical Investigations of Social Space*, edited by Jörg Blasius, Frédéric Lebaron, Brigitte Le Roux, and Andreas Schmitz, 345–57. Cham: Springer.

Myers, Zoë. 2020. *Wildness and Wellbeing: Nature, Neuroscience, and Urban Design*. Singapore: Springer Singapore.

Natale, Simone. 2021. *Deceitful Media: Artificial Intelligence and Social Life after the Turing Test*. New York: Oxford University Press.

Neumann, John von. (1958) 2012. *The Computer & the Brain*. 3rd ed. New Haven, CT: Yale University Press.

Nietzsche, Friedrich. (1872) 1999. *The Birth of Tragedy and Other Writings*. Edited by Raymond Geuss and Ronald Speirs. Cambridge Texts in the History of Philosophy. Cambridge: Cambridge University Press.

Nietzsche, Friedrich. (1887) 2007. *On the Genealogy of Morality*. Edited by Keith Ansell-Pearson. Translated by Carol Diethe. Cambridge: Cambridge University Press.

Ogihara, Mitsunori, and Jerry Bonnell. 2023. *Exploring Data Science with R and the Tidyverse: A Concise Introduction*. 1st ed. Boca Raton, FL: Chapman and Hall/CRC.

Oliveira, Thays A., Vitor N. Coelho, Helena Ramalhinho, and Miquel Oliver. 2019. 'Digital Cities and Emerging Technologies.' In *Smart and Digital Cities*, edited by Vitor Nazário Coelho, Igor Machado Coelho, Thays A. Oliveira, and Luiz Satoru Ochi, 197–207. Cham: Springer.

Patton, Paul. 2018. 'Philosophy and Control.' In *Control Culture: Foucault and Deleuze after Discipline*, edited by Frida Beckman, 193–210. Edinburgh: Edinburgh University Press.

Pavlenko, Vitaliy, and Tetiana Shamanina. 2023. 'Eye-Tracking Technology in Smart System for Monitoring of Human's Psychophysiological States.' In *Innovations in Smart Cities Applications*, edited by Mohamed Ben

Ahmed, Anouar Abdelhakim Boudhir, Domingos Santos, Rogerio Dionisio, and Nabil Benaya, 344–53. Cham: Springer.

Pearson, Karl. (1892) 1911. *The Grammar of Science*. 3rd ed. London: Adam and Charles Black.

Petty, William. (1682) 1888. *Essays on Mankind and Political Arithmetic*. Edited by Henry Morley. London: Cassell.

Phillips, William A. 2023. *The Cooperative Neuron*. Oxford: Oxford University Press.

Pietsch, Wolfgang. 2021. *Big Data*. Cambridge: Cambridge University Press.

Pietsch, Wolfgang. 2022. *On the Epistemology of Data Science: Conceptual Tools for a New Inductivism*. Cham: Springer.

Pitts-Taylor, Victoria. 2016. *The Brain's Body: Neuroscience and Corporeal Politics*. Durham, NC: Duke University Press.

Poole, David L., and Alan K. Mackworth. (2010) 2023. *Artificial Intelligence: Foundations of Computational Agents*. 3rd ed. Cambridge: Cambridge University Press.

Porter, Theodore M. (1986) 2020. *The Rise of Statistical Thinking: 1820–1900*. 2nd ed. Princeton, NJ: Princeton University Press.

Possati, Luca M. 2022. *Unconscious Networks: Philosophy, Psychoanalysis, and Artificial Intelligence*. London: Routledge.

Pykett, Jessica, Tess Osborne, and Bernd Resch. 2020. 'From Urban Stress to Neurourbanism: How Should We Research City Well-Being?' *Annals of the American Association of Geographers* 110 (6): 1936–51.

Rabinow, Paul, and Nikolas Rose. 2016. 'Biopower Today.' In *Biopower: Foucault and Beyond*, edited by Vernon W. Cisney and Nicolae Morar, 297–325. Chicago: University of Chicago Press.

Ray, William J. 2024. *Fundamentals of Brain and Behavior: An Introduction to Human Neuroscience*. New York, NY: Routledge.

Ricci, Franco, ed. 2006. *Difficult Games: A Reading of I Racconti by Italo Calvino*. Waterloo: Wilfrid Laurier University Press.

Risse, Mathias. 2023. *Political Theory of the Digital Age: Where Artificial Intelligence Might Take Us*. Cambridge: Cambridge University Press.

Rose, Nikolas. 2006. *The Politics of Life Itself: Biomedicine, Power, and Subjectivity in the Twenty-First Century*. Princeton, NJ: Princeton University Press.

Rose, Nikolas, and Joelle M. Abi-Rached. 2013. *Neuro: The New Brain Sciences and the Management of the Mind*. Princeton, NJ: Princeton University Press.

Rose, Nikolas, and Des Fitzgerald. 2022. *The Urban Brain: Mental Health in the Vital City*. Princeton, NJ: Princeton University Press.

Rosenblueth, Arturo. 1950. *The Transmission of Nerve Impulses at Neuroeffector Junctions and Peripheral Synapses*. Cambridge, MA: MIT Press.

Rouvroy, Antoinette, and Thomas Berns. 2013. 'Algorithmic Governmentality and Prospects of Emancipation.' Translated by Elizabeth Libbrecht. *La Découverte: Reseaux* 177 (1): 163–96.

Ruppert, Evelyn. 2012. 'Seeing Population: Census and Surveillance by Numbers.' In *Routledge International Handbook of Surveillance Studies*, edited by Kirsty Ball, Kevin Haggerty, and David Lyon, 209–16. London: Routledge.

Ruppert, Evelyn, and Stephan Scheel. 2019. 'The Politics of Method: Taming the New, Making Data Official.' *International Political Sociology* 13 (3): 233–52.

Ruppert, Evelyn, and Stephan Scheel. 2021. 'Data Practices.' In *Data Practices: Making up a European People*, edited by Evelyn Ruppert and Stephan Scheel, 29–48. London: Goldsmiths Press.

Russell, Stuart J., and Peter Norvig. 2022. *Artificial Intelligence: A Modern Approach*. 4th ed. Harlow: Pearson.

Schweber, Libby. 2006. *Disciplining Statistics: Demography and Vital Statistics in France and England, 1830–1885*. Durham, NC: Duke University Press.

Scott, James C. 2017. *Against the Grain: A Deep History of the Earliest States*. New Haven, CT: Yale University Press.

Scott, James C. (2017) 2019. *Homo domesticus: une histoire profonde des premiers États*. Translated by Jean-Paul Demoule and Marc Saint-Upéry. Paris: La Découverte.

Scott, James C. 2019. 'Response to Commentaries on Against the Grain.' *The Journal of Peasant Studies* 46 (4): 885–92.

Sejnowski, Terrence J. 2018. *The Deep Learning Revolution*. Cambridge, MA: MIT Press.

Shannon, Claude E. 1948. 'A Mathematical Theory of Communication.' *Bell System Technical Journal* 27 (3): 379–423.

Shapiro, Aaron. 2020. *Design, Control, Predict: Logistical Governance in the Smart City*. Minneapolis: University of Minnesota Press.

Shayan, Shadi, and Ki Pyung Kim. 2022. 'A Conceptual Framework to Manage Social Risks for Smart City Development Programs.' In *Resilient and Responsible Smart Cities*, edited by Hugo Rodrigues, Tomohiro Fukuda, and Simon Elias Bibri, 77–86. Cham: Springer.

Simmel, Georg. (1903) 1971. 'Metropolis and Mental Life.' In *On Individuality and Social Forms*, edited by Donald Nathan Levine, 324–39. Chicago: University of Chicago Press.

Simovici, Dan, and Kaixun Hua. 2019. 'Data Ultrametricity and Clusterability.' *Journal of Physics: Conference Series* 1334 (1): 1–11.

Sing, Robert, Tazuko Van Berkel, Robin Osborne, Robert Sing, Tazuko Van Berkel, and Robin Osborne. 2022. 'Doing Things with Numbers.' In *Numbers and Numeracy in the Greek Polis*, 1–23. Leiden: Brill.

Smail, Daniel Lord. 2008. *On Deep History and the Brain*. Berkeley: University of California Press.

Spicer, Zachary, and Austin Zwick. 2021. 'A Smart City for Toronto: What Does Quayside Tell Us about the State of Smart City Building?' In *The Platform Economy and the Smart City: Technology and the Transformation of Urban Policy*, edited by Austin Zwick and Zachary Spicer, 249–65. Montreal: McGill-Queen's University Press.

Spinoza, Benedictus de. 1985. *The Collected Works of Spinoza*. Edited by E. M. Curley. Vol. 1. 2 vols. Princeton, NJ: Princeton University Press.

Spinoza, Benedictus de. 2016. *The Collected Works of Spinoza*. Edited by E. M. Curley. Vol. 2. 2 vols. Princeton, NJ: Princeton University Press.

Sterratt, David, ed. 2011. *Principles of Computational Modelling in Neuroscience*. Cambridge: Cambridge University Press.

Stigler, Stephen M. 1999. *Statistics on the Table: The History of Statistical Concepts and Methods*. Cambridge, MA: Harvard Univ. Press.

Stoler, Ann Laura. 2016a. 'A Colonial Reading of Foucault: Bourgeois Bodies and Racial Selves.' In *Biopower: Foucault and Beyond*, edited by Vernon W. Cisney and Nicolae Morar, Chapter 16. Chicago: University of Chicago Press.

Stoler, Ann Laura. 2016b. *Duress: Imperial Durabilities in Our Times*. Durham, NC: Duke University Press.

Studley, Matthew E., and Hannah Little. 2021. 'Robots in Smart Cities.' In *How Smart Is Your City?* edited by Maria Isabel Aldinhas Ferreira, 75–88. Cham: Springer.

Sullivan, Gavin. 2022. 'Law, Technology, and Data-driven Security: *Infra*-legalities as Method Assemblage.' *Journal of Law and Society* 49 (S1).

Tashman, Adam P. 2024. *From Concepts to Code: Introduction to Data Science*. London: CRC Press.

Tazzioli, Martina. 2020. *The Making of Migration: The Biopolitics of Mobility at Europe's Borders*. London: Sage.

Teubner, Gunther, ed. 1988. *Autopoietic Law: A New Approach to Law and Society*. Berlin: de Gruyter.

Teubner, Gunther. 2019. *Critical Theory and Legal Autopoiesis: The Case for Societal Constitutionalism*. Edited by Diana Göbel. Manchester: Manchester University Press.

Thorvaldsen, Gunnar. 2018. *Censuses and Census Takers: A Global History*. London: Routledge.

Thurstone, Louis Leon, and E. J. Chave. 1929. *Measurement of Attitude*. Chicago: University of Chicago Press.

Tougaw, Jason Daniel. 2018. *The Elusive Brain: Literary Experiments in the Age of Neuroscience*. New Haven, CT: Yale University Press.

Trappenberg, Thomas P. 2022. *Fundamentals of Computational Neuroscience*. 3rd ed. Oxford: Oxford University Press.

Tryon, Robert Choate, and Daniel E. Bailey. (1939) 1970. *Cluster Analysis: Correlation Profile and Orthometric (Factor) Analysis for the Isolation of Unities in Mind and Personality*. New York, NY: McGraw-Hill.

Tufte, Edward R. (1983) 2001. *The Visual Display of Quantitative Information*. 2nd ed. Cheshire, CO: Graphics Press.

Turing, Alan. 1936. 'On Computable Numbers, with an Application to the Entscheidungsproblem.' *Proceedings of the London Mathematical Society* s2–42 (1): 230–65.

Turing, Alan. 1950. 'Computing Machinery and Intelligence.' *Mind: A Quarterly Review of Psychology and Philosophy*, New Series, 59 (236): 433–60.

Turing, Alan. (1948) 2004. 'Intelligent Machinery.' In *The Essential Turing: Seminal Writings in Computing, Logic, Philosophy, Artificial Intelligence, and Artificial Life, Plus the Secrets of Enigma*, edited by B. Jack Copeland, 410–32. Oxford: Oxford University Press.

Turing, Alan. (1951) 2004. 'Intelligent Machinery, A Heretical Theory.' In *The Essential Turing: Seminal Writings in Computing, Logic, Philosophy, Artificial Intelligence, and Artificial Life, Plus the Secrets of Enigma*, edited by B. Jack Copeland, 472–75. Oxford: Oxford University Press.

Turing, Alan. (1947) 2004. 'Lecture on the Automatic Computing Engine.' In *The Essential Turing: Seminal Writings in Computing, Logic, Philosophy, Artificial Intelligence, and Artificial Life, Plus the Secrets of Enigma*, edited by B. Jack Copeland, 378–94. Oxford: Oxford University Press.

Turing, Alan. 2004. *The Essential Turing: Seminal Writings in Computing, Logic, Philosophy, Artificial Intelligence, and Artificial Life, Plus the Secrets of Enigma*. Edited by B. Jack Copeland. Oxford: Oxford University Press.

Warwick, Kevin, and Huma Shah. 2016. *Turing's Imitation Game: Conversations with the Unknown*. Cambridge: Cambridge University Press.

Watson, Neil V., and S. Marc Breedlove. (2012) 2021. *The Mind's Machine: Foundations of Brain and Behavior*. 4th ed. Oxford: Oxford University Press.

Weber, Max. 1921. *The City*. Translated by Don Martindale. New York: Free Press.

Weber, Max. (1896) 1998. *The Agrarian Sociology of Ancient Civilizations*. London: Verso Books.

Wells, Robert V. 2015. *Population of the British Colonies in America Before 1776: A Survey of Census Data*. Princeton, NJ: Princeton University Press.

Wiener, Norbert. (1948) 1961. *Cybernetics: Or Control and Communication in the Animal and the Machine*. 2nd ed. Cambridge, MA: MIT Press.

Wiener, Norbert. (1950) 1989. *The Human Use of Human Beings: Cybernetics and Society*. London: Free Association.

Willis, Katharine S. 2020. 'The Death and Life of Smart Cities.' In *The Routledge Companion to Smart Cities*, edited by Katharine S. Willis and Alessandro Aurigi, 411–28. London: Routledge.

Wilson, Elizabeth Ann. 2010. *Affect and Artificial Intelligence*. Seattle, WA: University of Washington Press.

Yang, Dejun, Guoliang Xue, Xi Fang, and Jian Tang. 2016. 'Incentive Mechanisms for Crowdsensing: Crowdsourcing with Smartphones.' *IEEE/ACM Transactions on Networking* 24 (3): 1732–44.

Yeo, Eileen Janes. 2003. 'Social Surveys in the Eighteenth and Nineteenth Centuries.' In *The Modern Social Sciences*, edited by Theodore M. Porter and Dorothy Ross, 7:83–99. The Cambridge History of Science. Cambridge: Cambridge University Press.

Yoffee, Norman, ed. 2015. *The Cambridge World History. Volume 3, Early Cities in Comparative Perspective, 4000 BCE-1200 CE*. The Cambridge World History. Cambridge: Cambridge University Press.

Zanella, Andrea, Nicola Bui, Angelo Castellani, Lorenzo Vangelista, and Michele Zorzi. 2014. 'Internet of Things for Smart Cities.' *IEEE Internet of Things Journal* 1 (1): 22–32.

Zhang, Jianfeng, Xian-Sheng Hua, Jianqiang Huang, Xu Shen, Jingyuan Chen, Qin Zhou, Zhihang Fu, and Yiru Zhao. 2019. 'City Brain: Practice of Large-Scale Artificial Intelligence in the Real World.' *IET Smart Cities* 1 (1): 28–37.

Zhu, Mu. 2023. *Essential Statistics for Data Science: A Concise Crash Course*. Oxford: Oxford University Press.

Index

EU Authorised Representative:
Easy Access System Europe Mustamäe tee 50, 10621 Tallinn, Estonia
gpsr.requests@easproject.com

Printed and bound by CPI Group (UK) Ltd, Croydon, CR0 4YY
13/11/2025
01997198-0007